Detergents and Cleaners

Detergents and Cleaners

A Handbook for Formulators

Edited by K. Robert Lange

Hanser Publishers, Munich Vienna New York

Hanser/Gardner Publications, Inc., Cincinnati

The Editor:
Dr. K. Robert Lange, 805 Lombard Street, Philadelphia, PA 19147, USA

Distributed in the USA and in Canada by
Hanser/Gardner Publications, Inc.
6600 Clough Pike, Cincinnati, Ohio 45244-4090, USA
Fax: +1 (513) 527-8950

Distributed in all other countries by
Carl Hanser Verlag
Postfach 86 04 20, 81631 München, Germany
Fax: +49 (89) 98 12 64

The use of general descriptive names, trademarks, etc., in this publication, even if the former are not especially identified, is not to be taken as a sign that such names, as understood by the Trade Marks and Merchandise Marks Act, may accordingly be used freely by anyone.

While the advice and information in this book are believed to be true and accurate at the date of going to press, neither the authors nor the editors nor the publisher can accept any legal responsibility for any errors or omissions that may be made. The publisher makes no warranty, express or implied, with respect to the material contained herein.

Library of Congress Cataloging-in-Publication Data
Detergents and cleaners : a handbook for formulators / edited by K.
Robert Lange.
    p. cm.
Includes bibliographical references (p.   -) and index.
ISBN 1-56990-167-8
1. Detergents, Synthetic. 2. Cleaning compounds. I. Lange, K.
Robert.
TP992.5.D39   1994
668'.14—dc20   93-49047

Die Deutsche Bibliothek - CIP-Einheitsaufnahme
Detergents and cleaners : a handbook for formulators / ed. by
K. Robert Lange. - Munich ; Vienna ; New York : Hanser ;
Cincinnati : Hanser/Gardner, 1994
    ISBN 3-446-17307-2 (Hanser)
    ISBN 1-56990-167-8 (Hanser/Gardner)
NE: Lange, K. Robert [Hrsg.]

© Carl Hanser Verlag, Munich Vienna New York, 1994
Typeset in the USA by pageAbility, Willseyville
Printed and bound in Germany by Schoder Druck GmbH & Co. KG, Gersthofen

Preface

When first considering this task, I wanted to write an article on formulating, in general. Having spent many years formulating products and supervising chemists, my sympathy for the bench chemist's problems, and respect for the ingenuity shown, had led me to write a series of articles for my employer's house organ, explaining the field to the layperson. Editing and contributing to a book meant for the bench chemist is a different matter entirely. For this a focus is needed, one that will attract readers. The topic of detergency provides such a focus because it is of universal interest.

Whatever the industrial or home environment, people have a need for cleanliness. This includes their bodies, their clothing, and the equipment they use. Today there is a strong emphasis on keeping the entire environment preserved, and cleaners play their role in this effort. To these ends the chemical process industry develops formulated products, based on synthetic detergents, that are expected to attack soils of every imaginable type, regardless of water hardness or the conditions of use. Governments around the world monitor and regulate these efforts while encouraging cleanliness through campaigns and laws. Cleanliness is a major worldwide industry.

This book provides methodology for formulators generally, through the medium of detergency. Formulating principles are general, and this volume might just as well be on corrosion inhibitors or pie-baking; but the publisher produces *Tenside,* that fine journal devoted to detergency, and has a German-language handbook by Stache on that subject, among other things. Hanser's desire to produce a book on this topic for the North American market coincided with my own views of the appropriate field to use as the focus of a book for formulators.

The intended audience for this volume is the bench chemist or senior technician, working in a small to medium-sized laboratory, who wants to have a book of convenient size for daily reference and as a guide to the more theoretical, multivolume tomes in the library. My hope is that this book will fill a need between the two extremes of the theoretical texts and the fine trade magazines, which offer much useful information but can be difficult to use in terms of retrieval.

The authors of the various chapters have all overcome the daily pressures of their jobs to share their knowledge here. Working with them has been a pleasure, and I thank them for their efforts. I also thank the following for their cooperation: H.L. Williams, vice president at Monsanto; A.F. Hidalgo, vice president at Colgate Palmolive; and A.F. Burns, technical director at PQ, who were instrumental in recommending members of their staffs to author chapters.

K.R. Lange

Contents

Introduction

Every chemical product that reaches the ultimate consumer (with the possible exception of sugar) is, in some degree, a formulated product.* This being the case, it is somewhat mysterious to note the low esteem the formulation process possesses in both academia and industry. Certainly, no college department of chemistry or chemical engineering offers a course entitled *Formulation 101*. And, although formulation is often considered to be an arcane art, its practitioners are often referred to as "mere formulators." (One wonders, indeed, what the effect of calling a designer of perfumes and scents a "mere" perfumer would be!)

One of the reasons for this low esteem probably is the notion that formulation is largely Edisonian in nature and, therefore, does not possess the cachet that surrounds "real" research. First of all, it should be pointed out that there is nothing in principle wrong with the Edisonian approach to discovery, as long as it is applied with intelligence. Many important discoveries (even in basic science) have resulted from Edisonian methods, although this is often concealed by referring to serendipity.

A number of years ago, I presented a talk with the title *Philosophy of Formulation*.† Although, as the title of the symposium at which the talk was delivered suggests, my remarks were directed mostly to pesticide formulations, I made the point that formulation was *not* particularly Edisonian, but bore many similarities to what is commonly referred to as basic research.

On the other hand, formulation research is carried out for reasons different from those of the so-called basic research. Among those reasons might be listed the following:

* To render the use of application of the active component more effective.
* To enable the consumer to use the product more readily.
* To improve the stability of the product.
* Last—but not least—to improve the aesthetics of the product.

As I pointed out in my talk, to do it properly, one must plan the formulation study. This may involve answering a series of questions. For the formulation of pesticides, these questions were suggested:

* What is the active ingredient and what is its form?
* In what type of system will it be delivered (e.g., solution, dispersion, emulsion)?
* If a solution or dispersion, dissolved or dispersed in what?
* If an emulsion, o/w or w/o—and what is in the aqueous phase—what is the oil phase?

---

* Table salt, which one might think of as unformulated, is commonly "iodized" and additionally contains an anticaking compound.

† Becher, P. In *Pesticide Formulatins and Application Systems,* Vol. 8, D.A. Hovde and G.B. Beestman, Eds., American Society for Testing and Materials, Philadelphia, 1989, p. 13.

- How is it to be applied to the system: If a herbicide— dusted, sprayed? If a drug—applied topically, swallowed, injected?
- What is the desired composition (e.g., if an emulsion, what are the phase ratios)?
- If an emulsion or dispersion, what surface active agents should be used? What restrictions are imposed by safety considerations?
- What physical properties must the system have (e.g., viscosity, density)? In dispersions and emulsions, what is the effect of particle size distribution? If the system is viscoelastic, what effect does this have?
- How do we make it? What types of mixing or blending equipment? How about scale-up?
- Finally, how do we evaluate the system? What degree of stability is required? How is stability determined? What about quality assurance?

These questions can be restated to apply to any kind of formulation problem, and, no doubt, other questions will occur to the worker.

This brings us to an important point. It will not have escaped the eye of the reader that this book to a large extent emphasizes formulation of detergent species. This is simply a matter of convenience. From the foregoing, it should be obvious that the lessons and techniques discussed here can be transferred to any other field with only a modest effort. In other words, this book will be useful to *any* formulator, whether the problem at hand involves, among others, pharmaceuticals, cosmetics, paints, or—yes—pesticides. The authors and the editor are to be commended.

Paul Becher

CHAPTER 1

Basic Surfactant Concepts

Eric W. Kaler

This chapter introduces the fundamental ideas that guide the production of successful surfactant formulations. First the major features of surfactant molecules are discussed, and the kind of microstructures that form (micelles, vesicles, liquid crystals) reviewed. Next the general patterns of surfactant phase behavior are presented, and in particular the basic ideas needed to formulate microemulsions are summarized. The interfacial aspects of surfactants and their relation to spreading coefficients are reviewed, and the final section describes the basic properties of emulsions.

1.1   Introduction

Surfactant solutions exhibit many phenomena of scientific and technological interest. For example, surfactants are of primary importance in detergency, separation techniques, agriculture, and the pharmaceutical industry. Naturally occurring surfactants such as phospholipids are the main components of cell membranes and thus play a vital role in organizing the chemical reactions that sustain life. There are an enormous number of natural and synthetic surfactants available for applications, and choosing the optimal kind of surfactant for a given use can be a remarkable challenge.

Intelligent and rational use of surfactants in formulations for carrying hydrophobic materials such as drugs or pesticides into water, as well as in formulating cleaning products for textiles or hard surfaces, draws on principles from physical chemistry and thermodynamics. Moreover, the ultimate success or failure of a formula also often depends on the *kinetics* of solubilization or wetting. This introductory chapter describes a few of the most important properties of surfactants and shows in outline some of the unifying concepts in surfactant science. Useful references for further reading are given in the bibliography.

A surfactant (surface active agent) lowers the equilibrium interfacial tension between the medium in which it is dissolved and any other contacting fluid. Interfacial tensions are thermodynamic properties, like density and heat capacity, and so can be measured under given conditions and tabulated. If temperature and pressure are held constant, the value of the interfacial tension represents the work (and change in Gibbs free energy) necessary to create more interface. Since dispersing any material to colloidal size scales generates an enormous amount of interface, the ability of surfactants to reduce interfacial tensions is critical to such applications as the preparation of emulsions and the wetting and dispersal of powders in liquids. These interfacial properties of surfactants are discussed further in Section 1.6.

Eric W. Kaler, Center for Molecular and Engineering Thermodynamics, Department of Chemical Engineering, Colburn Laboratory, University of Delaware, Newark, Delaware 19716, U.S.A.

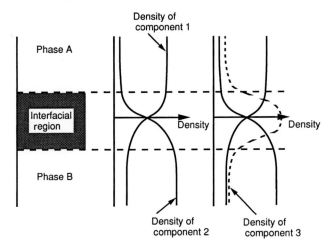

Figure 1.1 The relative densities of materials at the interface between two bulk liquids. Phase A has a high density of component 1, while phase B is rich in component 2. When an interfacially active material, component 3, is added, it accumulates in the interfacial region.

The term *surfactant* is also correctly applied to sparingly soluble substances that lower the interfacial tension between liquid phases by spreading spontaneously over their interface. In either case, the surfactant molecule is surface active, and in a dilute solution its concentration at the interface is higher than its concentration in the bulk phase (Fig. 1.1). For most applications of interest to the readers of this book, one of the phases will be water, and the other will be either a hydrophobic material (an "oil") or a solid surface. In these cases, the most effective surfactants not only lower interfacial or surface tensions, but also are amphiphilic. An amphiphilic molecule contains two parts: one is hydrophilic or "water-loving," and the other is lipophilic ("oil-loving") or hydrophobic ("water-hating"). The hydrophilic portion of the molecule (the "head group") is either an anionic, cationic, zwitterionic, or nonionic polar group. The hydrophobic "tail group" portion of the molecule is usually a single or double hydrocarbon chain with various degrees of unsaturation or substitution. Fluorosurfactants, in which all or portions of the hydrophobic moiety are fluorinated, can be particularly useful.

Restricting the standard definition of a surfactant to include only amphiphiles eliminates some surface active molecules (e.g., benzene) that do accumulate at the interface, but still includes those that may only slightly reduce interfacial tensions or promote the solubilization of oil and water. A still more restrictive definition is to include as surfactants only molecules that form microstructure in solution. Such a microstructure would include any of the micellar, liquid crystalline, or vesicle morphologies discussed below. This extended definition excludes amphiphiles such as small acids, alcohols, and amines that lower the surface tension of water but associate only weakly (as small oligomers) in solution. A corollary of this definition is that surfactants form aggregates with a distinct "inside" and "outside," with the boundary between the two regions corresponding to the connection between the hydrophilic and hydrophobic portions of the surfactant. The concept of a surfactant "sheet" that separates an oily region from an aqueous one is particularly useful in understanding the structures present in microemulsions and emulsions.

1.2   Surfactant Mechanisms

The mechanism through which surfactants act is directly related to the amphiphilic nature of these materials. The driving force for surfactant aggregation in solution is the so-called hydrophobic effect. The process of introducing an amphiphile into water is energetically favored because of the solubility of the head group, but the disruption of the hydrogen bonding structure of water by the tail group is unfavorable from an energetic point of view. This incompatibility of the tail and water is called the *hydrophobic effect*. Clearly the free energy of a solution of water and surfactant is minimized when the hydrophobic portion of the surfactant is expelled from water. When the surfactant is dilute, this can be accomplished by accommodating the surfactants at the interface between water and any other liquid or air. At higher amphiphile concentrations, there is another mechanism for shielding the tails from water, namely, the amphiphiles can aggregate in a cooperative way to form micelles or other solution structures such as vesicles or liquid crystals. The kinds of structures formed by surfactants and their uses are highlighted in the next section.

The kind of equilibrium microstructures formed by surfactants in solution depends sensitively on the geometry of the hydrophobic part of the molecule and the chemical nature of the hydrophilic portion. The sizes and shapes of the surfactant aggregates (and the transitions between them) also depend on the temperature and salinity or ionic strength of the solution, and can change dramatically as a hydrophobic material is solubilized. In practice, pure surfactants are difficult and expensive to produce, so homologous blends are usually used in formulations. Surfactant properties are often sensitive to small changes in the molecular structure, so the broadness of the molecular weight distribution in such blends is sometimes important to the success or reproducibility of a formulation. Broad mixtures of surfactants of different types are occasionally used in commercial formulations in an attempt to take advantage of different useful properties of some components, or to benefit from some synergism between pairs or complexes of different surfactants.

Surfactants are usually classified primarily by the chemical type of their hydrophilic part. A comprehensive list of surfactants and their main uses is given by Rosen, and periodical catalogues of commercial surfactants, such as the one by McCutcheon, are available. Anionic surfactants have negatively charged head groups; examples are sulfonates ($-SO_3^-$), sulfates ($-OSO_3^-$), and carboxylates ($-COO^-$), which are soaps. Anionic surfactants are often sold as ammonium or alkali metal salts. They are good solubilizers and are generally the cheapest choice for a given application.

Cationic surfactants have positively charged head groups. Common examples are the halide salts of long-chain primary amines ($-NH_3^+$) or quaternary ammonium ($-N(CH_3)_3^+$) ions. Primary amine surfactants are sensitive to pH and become ineffective at high pH. Cationic surfactants can bind strongly to negatively charged surfaces, and so find application as fabric softeners.

For both anionic and cationic surfactants, the typical hydrophobic group is a linear or branched hydrocarbon chain or chains. The chains range in length from about 8 to 20 carbons. Surfactants with shorter chains are too soluble in water and aggregate only weakly. Longer chains are frozen at moderate temperatures (and so do not easily form micelles); such surfactants are insoluble in water. Except near the chain melting temperature, the phase behavior of both anionic and cationic surfactants in water is only slightly affected by temperature.

Surfactant folklore teaches that anionic and cationic surfactants are "not compatible" inasmuch as the oppositely charged surfactant ions may pair and form an insoluble precipitate. While this certainly may happen, there is a rich and complex synergism between oppositely charge surfactant ions that can give rise to enhanced reductions in interfacial tension as well as facilitate the formation of unusual surfactant microstructures.

Nonionic surfactants have head groups containing no ionizable species; the typical example is the polyoxyethylene group, but there are many others. Plant-based surfactants with sugar head groups are nonionic. Nonionic surfactants are usually liquids, as opposed to powders or pastes. In water, nonionic surfactants are less sensitive to the presence of electrolytes than are ionic surfactants, and so are good for applications with hard water. Aqueous solutions of polyoxyethylene surfactants display unusual phase behavior with temperature. Usually surfactants become more miscible with water as temperature increases, but an aqueous phase containing nonionic surfactant will separate upon heating into two phases, one surfactant-rich and the other surfactant-poor. The homologous series of polyoxyethylene nonionic surfactants of the type $[C_iH_{2i+1}]$–$[OCH_2CH_2]_j$–OH (so-called C_iE_j) are particularly convenient for fundamental research in surfactant phase behavior because the relative balance between the hydrophobic and hydrophilic portions of the molecule can be tuned in small discrete steps by changing i and j. Practical polyoxyethylene nonionic surfactants are the Neodol (Shell) and Brij (ICI) families.

The final major class of surfactants have zwitterionic hydrophilic groups. The most important examples are lecithins—double-chained surfactants with phosphatidylcholine head groups $[RCH_2(PO_4)^-CH_2CH_2N^+(CH_3)_3]$—which are major components of biological membranes. The double hydrocarbon chains are linear with between 12 and 20 carbons each and usually have some unsaturation. Betaines $[RN^+(CH_3)_2CH_2COO^-]$ are examples of synthetic zwitterionics. Many zwitterionic surfactants behave like cationic surfactants at low pH and anionic surfactants at high pH. The ability to switch the nature of the surfactant by changing a solution property provides a potentially useful way to control microstructures or wetting properties. At intermediate pH, zwitterionics are effectively uncharged and, like nonionic surfactants, are not very sensitive to water hardness.

More details about surfactants can be found in Chapter 3. Suppliers are listed in the Appendix.

1.3   Surfactant Microstructures

All surfactant molecules have some solubility as monomers in water, although it may be vanishingly small. As the overall surfactant concentration increases, the hydrophobic effect can drive surfactant monomers to form microstructures—aggregates of colloidal dimensions—that exist in equilibrium with the molecules or ions from which they form. The concentration at which these microstructures form is called the critical aggregation concentration (CAC). At bulk concentrations above this critical value, surfactant is found in solution either as monomer (at a concentration roughly corresponding to the critical aggregation concentration) or associated with one or more kinds of surfactant microstructure. The distribution of surfactant between monomer and aggregates is shown schematically in Fig. 1.2.

The dramatic change in the state of surfactant aggregation around the CAC means that many measurable properties of the solution also show more or less sharp changes at the CAC (Fig. 1.3). For example, Fig. 1.3 illustrates the changes in turbidity caused by the presence of small aggregates above the CAC. If the surfactant is ionic, the electrical conductivity of the solution will also change because the larger aggregates will be less effective charge

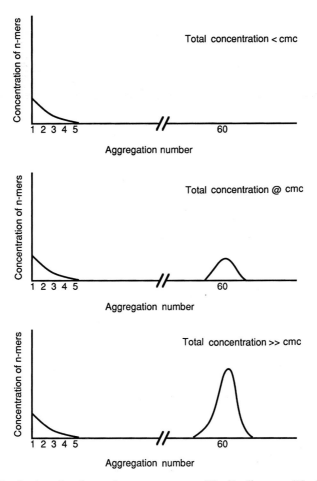

Figure 1.2 The distribution of surfactant between monomer ($N = 1$), oligomers ($N = 2, 3, 4, \ldots$) and micelles ($N = 60$) as a function of total surfactant concentration.

carriers than the smaller and more mobile monomers. The behavior of the surface tension above the CAC reflects the fact that the addition of more surfactant above the CAC produces more aggregates, while the monomer and surface concentrations remain nearly constant. The increase in dye solubility above the CAC is due to the uptake of the hydrophobic dye into the hydrophobic region of the aggregate. Many other more sophisticated techniques, such as NMR or fluorescence spectroscopy, also provide enough information at the molecular level to permit one to detect changes in aggregation.

For practical laboratory use, surface tension measurement is the most convenient and suitable method to determine the critical micelle concentration (cmc) of any surfactant. The surface tension plot also provides a quick measure of the purity of the surfactant. Contaminants usually cause the surface tension in the vicinity of the cmc to drop below its ultimate value, so instead of the sharp break in the curve as shown in Fig. 1.3, there is a "dip" or local minimum. For ionic surfactants, electrical conductivity is also a convenient and reliable way to determine cmc values.

The kind of aggregate that forms depends on a number of properties of the surfactant and the solution. We will first consider the simplest kind of microstructure, the micelle.

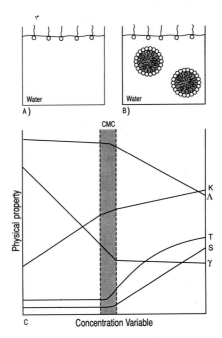

Figure 1.3 Changes in surfactant solution properties as a function of surfactant concentration: (A) the surfactant at the interface when the total surfactant concentration is below the cmc, (B) the saturated interface and micelles above the cmc, and (C) the characteristic variation in physical properties around the shaded cmc region. The properties are electrical conductivity K, specific conductance Λ, turbidity T, solubility S, and interfacial tension γ.

1.3.1   Micelles

Micellization is a cooperative process in which a large number of monomers (from 50 to 100 in a spherical micelle) assemble to form a closed aggregate in which the nonpolar parts of the amphiphiles are, on average, shielded from water. When micelles form, the CAC is called the critical micelle concentration (cmc).

The cmc depends sensitively on the number, length, and nature (saturation, branching, or substitution) of the hydrophobic chains and on the type of head group. Representative cmc values are given in Table 1.1. For a given hydrophobic tail (e.g., C_{12}), the cmc of a nonionic surfactant is generally much lower than that of an ionic. Values of cmc change by only about a factor of two for ionic surfactants of a given tail length for a variety of head groups and counterions.

Several correlations are available for estimating cmc values. For hydrogenated surfactants with a single unbranched saturated tail containing n_c carbon atoms, cmc data are well correlated by the expression

$$\log_{10} (\text{cmc}) = A - Bn_c \qquad (1.1)$$

where A and B are temperature-dependent constants for a given type of surfactant (Table 1.2). Added electrolyte suppresses the cmc's of both ionic and nonionic surfactants, although the change is much larger for ionics. Useful correlations are of the form $\log_{10}(\text{cmc}) = A_1 + B_1 \log C$ for ionic surfactants and $\log_{10} (\text{cmc}) = A_2 + B_2 C$ for nonionics, where C is the electrolyte concentration. There are more subtle dependencies on the type and valence of electrolyte,

Table 1.1   Representative Critical Micelle Concentrations

Surfactant	Cmc (*M*)
$C_{12}SO_4^-Na^+$	8.1×10^{-3}
$C_{12}SO_3^-Na^+$	1.0×10^{-2}
$C_{12}CO_2^-K^+$	1.2×10^{-2}
$C_{12}E_4$ (25 °C)	4.6×10^{-5}
$C_{12}E_5$ (25 °C)	6.4×10^{-5}
$C_{12}E_6$ (25 °C)	8.7×10^{-5}
$C_{12}N(CH_3)_3^+Br^-$	1.5×10^{-2}
$C_{16}N(CH_3)_3^+Br^-$	8.2×10^{-4}
$(C_{16}N(CH_3)_3)_2^+SO_4^-$	6.0×10^{-4}

Source: Data selected from Refs. [1] and [9].

Table 1.2   Correlative Parameters Describing the Dependence of the cmc (*M*) of Single-Tailed Surfactants on the Number of Carbons (*n*) in the Surfactant Tail: $\log_{10}(cmc) = A - Bn$.

Surfactant type	*T* °C	*A*	*B*
$C_n CO_2^-Na^+$	20	2.41	0.341
$C_n CO_2^-K^+$	25	1.92	0.290
$C_n SO_3^-$	40	1.59	0.294
$C_n OSO_3^-$	45	1.42	0.295

Source: Ref. [10].

with di- and trivalent electrolytes being much more effective than monovalent ones in reducing critical micelle concentrations.

Altering the architecture of the hydrophobic portion of the surfactant changes the cmc. Changes that interfere with chain packing in the core of the micelle generally increase the cmc. Examples are unsaturation, branching, and the addition of most polar groups to the chain. Adding a benzene ring is equivalent to adding three carbons to the chain. Changes of these kinds also affect the packing in the analogous liquid hydrocarbons and are reflected in the melting temperatures of the hydrocarbons. Perfluorination of the chain lowers the cmc dramatically, and a rule of thumb is that a perfluorinated chain containing *n* carbons behaves like a hydrogenated chain of length *n* + 3.

The tendency of a surfactant to aggregate, and its cmc, can also be changed by the addition of organic molecules. Amphiphilic molecules that are only sparingly soluble in water (e.g., higher alcohols) tend to participate in micelle formation by intercalating into the surfactant layer at the micelle surface. They thus tend to lower the cmc, particularly for ionic surfactants. Low concentrations of alcohol (on the order of the cmc) are sufficient to change the cmc noticeably.

Other water-soluble organic molecules act to change the properties of bulk water. These materials are classed phenomenologically either as "structure makers," which decrease the cmc of surfactants, or "structure breakers," which increase the cmc. Xylose and fructose are structure makers, while urea and formamide are examples of structure breakers (these points are discussed extensively in the text by Hunter listed in the bibliography). Clearly the presence of such materials at substantial concentrations changes the bulk properties of water (e.g., its dielectric constant) and also interferes with formation of the characteristic hydrogen bond network of water. The fundamental mode of action of these molecules is not known.

Micelles form in many other hydrogen bonding solvents besides water. Polar organic molecules like glycerol, propylene glycol, and formamide form hydrogen bonds, but such bonds generally do not generate the three-dimensional structure characteristic of hydrogen bonds in water. Micelle (and other microstructure) formation has been observed in fused salts such as ethylammonium nitrate (EAN), which does form a three-dimensional hydrogen bond network. Study of the mechanism of micelle formation in exotic solvents provides a valuable test of the generality of the ideas about "hydrophobic" interactions but also has important practical applications. Nonaqueous formulations are useful for products as diverse as antifreeze and skin lotion, and surfactant structures in fused salts can be useful for electro-chemical reactions.

The polar organic molecules mentioned above are similar to water in that they have high dielectric constants and are immiscible with hydrocarbon oils. In general, the cmc of ionic and nonionic surfactants is higher in nonaqueous media than in water. The formation of the liquid crystalline structures discussed below is also suppressed in polar organic solvents. In EAN, the cmc's are also larger than those in water, and the micelles that eventually do form are smaller than those of the same surfactant in water. Neat EAN has an ionic strength of about 12 M, so ionic interactions are nearly completely suppressed.

The cmc and micelle behavior depends on the solvent–oil interfacial tension in a complicated way. Nonetheless, the lower interfacial tensions between polar organics and

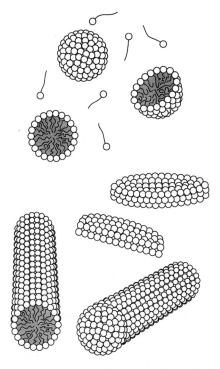

Figure 1.4 Cartoons of micelle structures. Modified from the original drawings of F.B. Rosevear [J. Soc. Cosmet. Chem., 19, 581 (1968)].

water (ca. 20–30 dynes/cm vs. 50 dynes/cm at the interface between water and oil) means that there can be substantially more contact between the surfactant tails and a polar organic than between the tails and water for the same cost in energy. Measured values of the area occupied by a surfactant molecule in an aggregate in a polar organic solvent are nearly twice those measured in water. Thus the structures in polar organic solvents are more open and feature substantial contact between the micelle core and the solvent.

At surfactant concentrations only slightly above the cmc range, the micelles formed from most surfactants in water are roughly spherical. However, as either the surfactant concentration or the concentrations of an electrolyte such as NaCl is increased, the micelle shape can become either prolate ellipsoidal or even cylindrical (Fig. 1.4). The formation of oblate or disklike micelles has been reported in solutions of surfactants of several diverse kinds, but further examination has always shown the micelles to be in fact prolate, or the structures to not be micelles at all. Nonetheless, disklike micelles probably do form in *mixtures* of surfactants.

1.3.2   Liquid Crystals

The delicate balance between the nature of the head and tail groups of a surfactant can lead to micelle formation at a particular set of concentration and temperature conditions. For many surfactants, however, the molecular geometry is such that the surfactant cannot pack successfully into a closed micelle because, for example, its tail group is too bulky or branched, or its head group is too small to shield the necessary core volume. In any case, the surfactant is still forced to comply with the hydrophobic forces and aggregate in some way to remove the tails from water. Sometimes the thermodynamically preferred form of aggregation is that of a liquid crystal.

Liquid crystals are microstructures in which molecules are able to move locally as a fluid but have long-range order in one, two, or three dimensions. As a pure solid is heated, it most often changes from an ordered or amorphous solid to an isotropic liquid. There are, however, materials that do not follow this simple pattern, but instead display intermediate thermodynamically stable phases with various kinds of long-range order. These temperature-induced (thermotropic) liquid crystalline phases or mesophases can also change state in response to external fields, and are widely applied.

Although pure surfactants may demonstrate thermotropic behavior at higher temperatures, lyotropic ("solvent-induced") liquid crystal phases can form as water or oil is added to surfactant. These lyotropic liquid crystals form directly as a result of the amphiphilic nature of the surfactant. For surfactants that form micelles, lyotropic mesophases are typically found at surfactant concentrations higher than those of micellar solutions. The most common mesophase morphologies are the hexagonal phase (H), which is assembled of aggregates similar to rodlike micelles, and the lamellar phase (L_α) (Fig. 1.5). It is interesting that both micellar and liquid crystal phases are topologically ordered by the surfactant—that is, the surfactant imposes an "inside" and an "outside" on both structures—but the liquid crystalline phase has in addition long-range geometric order.

Some surfactants that do not form micelles at all in water are able to form liquid crystals (usually lamellar) at very low surfactant concentrations. Phospholipids or other double-tailed surfactants almost always form lamellar phases as their first state of aggregation in water. Detecting such submicroscopic liquid crystals is experimentally difficult, particularly

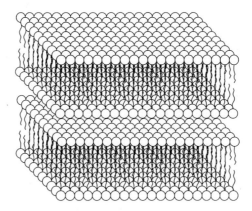

Figure 1.5 Schematic structure of the lamellar phase L$_\alpha$.

because most of the physical properties of the dispersion will change in much the same way as shown in Fig. 1.3 regardless of whether micelles or liquid crystals form. If the surfactant initially forms a liquid crystal phase, then the liquid crystals and water are properly called a two-phase dispersion, and the critical aggregation concentration identified by the changes in physical properties is the solubility limit of the surfactant.

1.3.3   Vesicles

The cartoon of the lamellar phase in Fig. 1.5 is a little misleading, for clearly the surfactant will try to organize to protect the edges of the layers from water. For this reason, dispersing a surfactant that forms lamellar liquids in water does not produce infinitely large surfactant sheets, but rather a dispersion of roughly spherical structures called liposomes. The bilayers in a liposome are arranged like the layers of an onion. Liposomes formed by gently shaking surfactant in water are usually more than a micrometer in size, so the dispersions are turbid. The internal bilayer structure of the liposomes causes them to be optically active, so they are also easily identified by a characteristic "Maltese cross" appearance in a polarizing light microscope.

Prolonged ultrasonication of a liposome dispersion produces a transparent dispersion of submicroscopic spheroidal vesicles (Fig. 1.6). Such small structures can be visualized only by electron microscopy or video-enhanced optical microscopy. Vesicles are usually classified as multilamellar vesicles (MLV), large unilamellar vesicles (LUV) with diameters above 0.1 μm, or small unilamellar vesicles (SUV). It is possible to prepare vesicles by treating liposome dispersions in other ways (e.g., ultrafiltration or microfluidization) or by solvent or detergent extraction of micellar or other solutions. Vesicles are obviously attractive as delivery agents for pharmaceutical molecules or other active materials that could be trapped in the vesicle core.

The thermodynamic stability of vesicles prepared by treatment of liposome dispersions is still a matter of some debate. Long-term observations of phospholipid vesicles are hampered by chemical and biological degradation of the surfactant, but observations of vesicles made of other surfactants points to the slow reversion of vesicles to the large liposomes from which they came. Such reversion, of course, limits the shelf life of a vesicle

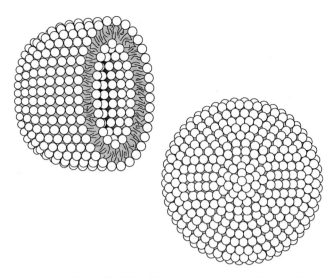

Figure 1.6 Schematic view of a small unilamellar surfactant vesicle. Such vesicles are typically about 500 Å in diameter.

formulation, but, as is the case for emulsion formulations, clever engineering can produce a usable product.

Recently, a second class of vesicle structures has been reported. In this case, the vesicles are observed to form *spontaneously* when certain surfactant mixtures are added to water. These vesicles form without the addition of energy and are stable indefinitely. In each case reported, a mixture of surfactants (or at least a mixture of counterions with a single kind of surfactant ion) is required for the vesicles to form. The most extensive study to date describes vesicles formed by mixtures of anionic and cationic surfactants. In this case, it is likely that ion pairs form with the properties necessary to support bilayer formation, while excess unpaired surfactant acts to fluidize the membrane and promote its curvature.

1.3.4   Inverted Structures

Some surfactants have head groups that are too small, or tail groups that are too large, to successfully pack into bilayers. Such amphiphiles often form inverted structures when dispersed in a hydrophobic solvent. In these structures, the head groups are clustered together and the tails are arranged radially into the solvent. The head groups are usually solvated by water to some degree, so the core of an inverse micelle contains a small water pool.

The formation of such inverted micelles can be described thermodynamically in the same way as the formation of normal micelles; in other words, there is a change in the surface or bulk properties of the solution at some critical concentration (as in Fig. 1.3). However, the driving force for inverse micelle formation (the exclusion of head groups from solvent) is not as specific as the hydrophobic effect, and as a result the onset of inverted micelle formation is more gradual than that observed for normal micelle formation.

Inverted structures have a range of uses. The micelles usually have some capacity to take up additional water into their cores, and ultimately they form water-in-oil microemulsions as discussed below. This ability to create tiny, isolated water-rich regions makes such solutions interesting reaction media. Hydrophilic material can also be loaded into the micellar core and so solubilized in oil; such structures are used in engine oil additives and hydraulic and cutting oils. Inverse micelles are also useful for the separation of biological molecules (proteins) because the partitioning of biomolecules can be adjusted, for example, by changing the size of the micelle, by changing the concentration of surfactant or water, or by changing the temperature. The most studied inverted micellar phases are those formed by the double-tailed anionic surfactant Aerosol-OT [sodium bis (2-ethylhexyl)sulfosuccinate] in solvents like heptane or isooctane.

1.3.5   Other Structures

So far we have described relatively simple structures with straightforward Euclidean shapes. The organization of surfactants in solution is not always so simple, and there have been many reports of surfactant structures in which the surfactant is organized in more complicated ways. The most interesting are those in which the surfactant sheet is multiply connected, as shown in Fig. 1.7. This structure is schematically that of the so-called L_3 phase, which is formed by many surfactants.

The L_3 phase is made up of a surfactant bilayer that is everywhere shaped like a saddle. The sheet separates the water into two interpenetrating but separate volumes. Even though this kind of structure seems esoteric, research on L_3 phases formed with surfactants that contain a polymerizable group suggests that it might be possible to polymerize the structure, thereby forming unique porous materials.

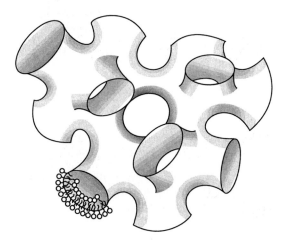

Figure 1.7 Artist's image of the microstructure in the L_3 phase. The surface represents the midplane of the surfactant bilayer sheet, which is shown partially in cross section on only one of the "arms." The surfactant sheet is everywhere saddle-shaped. Modified from the original sketch of Refs. [6] and [7].

1.3.6   Molecular Packing

This chapter has provided a qualitative description of the kinds of microstructure most commonly seen when surfactant is added to a solvent. From these descriptions emerge several clear and important trends that can be organized in terms of the molecular architecture of the surfactant. These trends were discussed first by C. Tanford and then made more quantitative by J. Israelachvili and colleagues (see Section 1.8). The following observations apply only to dilute surfactant solutions. When the concentration of surfactant aggregates is high enough that aggregate–aggregate interactions are important, then intra-aggregate packing is only one of several factors that set the equilibrium structure.

The basic idea is simple. For a surfactant microstructure to form (and be stable), the interior of the aggregate must contain all the hydrophobic portions of the surfactant molecules, and at the same time the surface of the aggregate must be covered by the hydrophilic heads. On the other hand, the details of the forces setting the packing of chains in the hydrophobic core, and, especially, the forces governing how the heads are solvated, are of course far from simple. Some progress in understanding the formation of structure can be made, however, by recognizing that the forces at work will set effective properties of the surfactant that can be described in terms of three geometric properties of any surfactant molecule. These three properties are the length of the hydrophobic chain l_c, the volume of the hydrophobic group v, and the area subtended by the head group a. This is illustrated in Fig. 1.8 for two common kinds of surfactant. Both surfactants have approximately the same area per head group and chain length, but the hydrophobic part of the double-tailed surfactant is considerably more bulky.

For surfactants to form a spherical micelle of radius R_m and aggregation number N, clearly the volume and surface area constraints on the micelle are:

$$Nv = \frac{4}{3}\pi R_m^3 \text{ and } Na = 4\pi R_m^2 \tag{1.2}$$

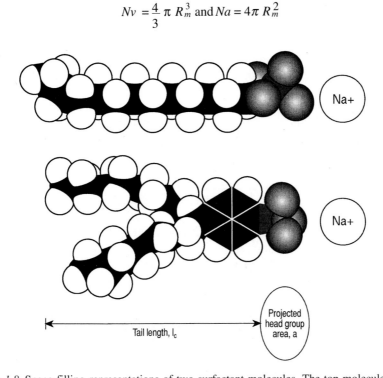

Figure 1.8 Space-filling representations of two surfactant molecules. The top molecule is sodium dodecyl sulfate (SDS), which has a single hydrophobic tail. The bottom molecule is sodium 4-(1´heptylnonyl) benzenesulfonate (SHBS), which has a branched hydrophobic tail that is considerably more bulky than the straight alkane tail of SDS. Also shown are the length of each surfactant and the area per head group. These parameters are needed for calculation of the surfactant number.

Table 1.3   Surfactant Packing Parameters for Different
Aggregate Geometries

Range of v/al_c	Predicted structure
$\leq \frac{1}{3}$	Spherical micelles
$\frac{1}{3}-\frac{1}{2}$	Cylindrical micelles
$\frac{1}{2}-1$	Bilayers and vesicles
≥ 1	Inverted structures

The ratio of these two equations gives $v/a = R_m/3$, and, since at least one of the hydrophobic chains must reach to the center of the micelle, $R_m \leq l_c$. Thus

$$\frac{v}{al_c} \leq \frac{1}{3} \qquad (1.3)$$

is the packing criterion for the formation of a spherical micelle. The ratio v/al_c is dimensionless and is sometimes called the surfactant number. Development of surfactant number criteria for other structures is easily done, and the results are presented in Table 1.3.

The surfactant number does describe the behavior of real surfactants. For example, for the anionic surfactant sodium dodecyl sulfate (SDS) in water, one can estimate the volume of the hydrophobic tail group to be 350 $Å^3$ and of the length of the hydrocarbon chain to be about 16.7 Å. The area per head group is difficult to estimate because of the role of electrostatic interactions, but experiment shows that the value above the cmc is 60 $Å^2$. Thus the SDS surfactant number is roughly 0.35, which is slightly above 1:3 and is consistent with the experimental observation of roughly spherical micelles in dilute solutions of SDS.

Such agreement is not completely a numerological accident, but surfactant number arguments are too crude to be relied on for quantitative guidance. Nonetheless, they are useful in shaping good intuitive thinking about how surfactants behave in solution.

Many trends in the evolution of surfactant microstructures with changes in temperature or ionic strength can be rationalized in terms of the behavior of the surfactant number. For example, adding electrolyte to a solution of spherical micelles of an ionic surfactant usually promotes the elongation and growth of the spherical micelles into cylindrical or threadlike micelles. This can be interpreted as a result of the electrolyte screening the electrostatic interactions of the head groups, thereby reducing the effective value of a and driving the surfactant number above 1:3.

The surfactant number idea also can serve as a guide to the evolution of microstructures formed by mixtures of surfactants. An example is the combination of the double-tailed surfactant didodecyldimethylammonium bromide (DDAB) and its single-tailed analog, dodecyltrimethylammonium bromide (DTAB). When dispersed in water at low concentrations, DDAB forms lamellar phases, while DTAB forms small spherical micelles. As DTAB is added to a DDAB lamellar phase, the observed progression of structure is from lamellae to equilibrium vesicles to rods to micelles as the effective value of v/al_c falls from more than 1:2 to less than 1:3.

1.4   Surfactant Phase Diagrams

A formulation chemist is often faced with choices about the kind and concentration of surfactant to use in a particular product. To have some rational structure for making such decisions, it is necessary not only to understand the kinds of microstructure that form but also to realize where the various microstructured phases are with respect to each other on temperature–concentration phase diagrams. Partial or schematic phase diagrams are available for many surfactants in the technical literature, and Figs. 1.9 and 1.10 are sketches of the main features of phase diagrams for ionic and nonionic surfactants, respectively.

The ionic surfactant phase diagram shows the location of the various phases as a function of surfactant concentration and temperature. The shaded markings denote two-phase regions where, for example, micellar and liquid crystalline phases coexist. At low temperatures, the surfactant is below its solubility or Krafft boundary. In this case, the surfactant is ineffective in forming microstructure or in changing interfacial tensions. While the surfactant may be slightly soluble in water at low temperature, higher surfactant concentrations are insoluble and exist as hydrated crystals in equilibrium with excess water. The Krafft boundary depends on the length of the surfactant tail group; for example, it is less than 10 °C for sodium dodecyl sulfate and is roughly 30 °C for sodium cetyl sulfate.

At temperatures above the Krafft temperature, the surfactant becomes dramatically more soluble in water. Now as surfactant is added, in this diagram, the first structural transition occurs at the critical micelle concentration. For ionic surfactants, the cmc increases slightly with temperature. As discussed above, depending on the molecular architecture of the surfactant monomer, either spherical or cylindrical micelles can form at the cmc. Even if spherical micelles are the first state of aggregation, there is often an evolution to cylindrical micelles as the surfactant concentration increases. The details of this transition depend strongly on the particular surfactant and the presence of additives (e.g., electrolytes). The most important physical manifestation of micelle growth into cylinders or rods is a significant increase in viscosity.

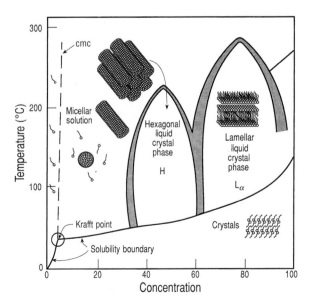

Figure 1.9 Schematic phase diagram for an ionic surfactant, showing the relative location of the liquid crystal phases and the low temperature solubility or Krafft boundary. The region around the cmc is shown at artificially high concentrations for clarity; on this concentration scale it would normally be indistinguishable from the temperature axis.

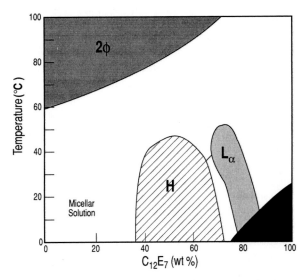

Figure 1.10 The phase diagram for water and the nonionic surfactant $C_{12}E_7$ as a function of temperature and surfactant concentration. Note the "cloud point" phase separation at high temperatures and low surfactant concentrations. Figure redrawn from Ref. [12].

Again in this schematic view of phase behavior, the next state of aggregation is the first liquid crystalline phase, which is usually one with hexagonal symmetry. The elements of the hexagonal phase are the cylindrical micelles found in the adjacent micellar phase, but the hexagonal elements are quite long. As surfactant concentration continues to increase, the hexagonal phase gives way to a lamellar phase, which in turn is bounded by a hydrated surfactant crystal phase at the surfactant boundary of the diagram. In all cases, the ordered phases melt at higher temperatures.

The phase behavior of nonionic surfactants in water at low temperatures is similar to that of ionic surfactants. On the other hand, many nonionics behave in dramatically different ways at higher temperatures (Fig. 1.10). Some nonionic surfactants become less soluble in water as temperature increases, and in fact separate into surfactant-rich and surfactant-lean phases at high temperatures. This "retrograde" phase behavior is unusual because most materials are more soluble in water at high temperatures than at low temperatures. The high temperature phase behavior is characterized by a lower consolute or critical point, and a lower consolute curve. Because the onset of phase separation is obvious visually as an increase in turbidity of the sample, this phase boundary is often called the cloud curve, and the lowest point on the curve is properly called the cloud point. The lowest point on the curve is at a low, but not zero, concentration, although the concentration scale in Fig. 1.10 does not allow it to be seen. The same temperature effect also causes the cmc of nonionic surfactants to decrease slightly with temperature, rather than increase as is the case for ionic surfactants.

The location of the cloud curve, both in temperature and concentration, depends on the molecular weight and structure of the particular nonionic surfactant, and it can be moved by the addition of electrolyte (Table 1.4). The pattern of phase behavior can be seen most clearly in the family of the C_iE_j polyoxyethylene nonionic surfactants as i and j change. The small amphiphile C_4E_1 has a cloud point at a temperature of 44.5 °C and a concentration of 27 wt %, and its surfactant character is too weak for liquid crystalline phases to form. As the surfactant is made longer, the cloud temperature drops and liquid crystal phases form. The phase

Table 1.4   Cloud Point Temperatures and Concentrations of
Various Nonionic Surfactants

Surfactant	T_c (°C)	C_c (wt %)
C_4E_1	44.5	27
C_6E_3	45.4	13
C_8E_3	8	4
$C_{10}E_3$	44.5	27
$C_{10}E_5$	40	3.5
$C_{10}E_6$	60	
$C_{12}E_5$	26	1.0
$C_{12}E_6$	48	2.2
$C_{12}E_7$	59[1]	
$C_{12}E_{12}$	>100	

[1] From Ref. [12].

Source: Except as noted, data from Ref. [11].

diagram for $C_{12}E_7$ shows both the cloud curve and liquid crystal phases (Fig. 1.10). For this surfactant, the cloud curve region is well separated from the liquid crystal phases. If the hydrophilic part of the surfactant is made shorter, for example to $C_{12}E_5$, then both the high temperature two-phase region and the liquid crystal phases grow on the phase diagram, and eventually they "collide" to produce new two-phase regions (Fig. 1.11). Along with this even more complicated phase behavior comes the appearance of the L_3 phase at temperatures above the cloud temperature and at surfactant concentrations below the "nose" of the L_α phase. The structure of this interesting bilayer phase is described above.

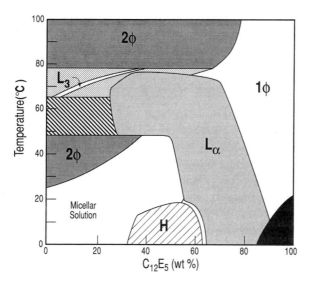

Figure 1.11 The phase diagram for water and the nonionic surfactant $C_{12}E_5$ as a function of temperature and surfactant concentration. The cloud point curve and the liquid crystal phases now intersect, and there are several new two-phase regions as well as a L_3 phase. Figure redrawn from Ref. [12].

1.5   Microemulsions

Of course, surfactant alone in water is of limited use in designing a formulation. Of much more utility are surfactant-containing solutions that solubilize oil into a mostly aqueous solution, or vice versa. In a broad sense, solubilization of water and oil together can be done in one of two ways. The first way is by forming an emulsion, in which two distinct thermodynamic phases (usually oil and water) are made kinetically stable by the addition of surfactant and the application of appropriate mechanical shear or mixing. The properties of emulsions are discussed below in Section 1.7. The second way is by formation of a *micro*emulsion, which is a single equilibrium phase containing at least water, oil, and surfactant. Microemulsions are thermodynamically stable phases, and so are fundamentally different from emulsions.

Microemulsion formulations offer some substantial advantages over emulsions. For example, the characteristic size in a microemulsion can be as small as 100 Å, so many microemulsions are either transparent or only slightly turbid. Microemulsions form without the input of mechanical energy, are of low viscosity, have low interfacial tensions with many materials, and are stable. Many additives can also be solubilized by microemulsions. The technique of microemulsion polymerization provides new and useful polymers. Applications of microemulsions have been limited in the past by a lack of appropriate scientific knowledge, but there is presently no reason to avoid their use as part of a process of rational formulation design.

Microemulsions are especially interesting and useful arrangements of material. Understanding of their fundamental properties, microstructures, and phase behavior is still evolving, but several aspects are well established. We will first review the major kinds of structure found in microemulsions and then explore some elementary aspects of microemulsion phase behavior.

The most obvious microemulsion structures are those that evolve from simple micelle phases upon the addition of, for example, oil to a normal micellar solution. The resulting swollen micelle or microemulsion droplet is illustrated in Fig. 1.12A. Such solutions are called oil-in-water (o/w) microemulsions. The basic structure is one of an oil core surrounded by a surfactant-rich sheet. The experimental evidence is such that droplets are approximately spherical, or are at most slightly prolate ellipsoids (American footballs) with aspect ratios of less than three. At the other extreme, addition of water to a solution containing inverted micelles again swells the cores, and swollen inverted micelles or water-in-oil (w/o) microemulsion droplets form (Fig. 1.12C).

In both cases, the minority component of the microemulsion (either water or oil) is enclosed by a surfactant-rich sheet, and that sheet is curved either toward the water (in a water-in-oil structure) or toward the oil (in an oil-in-water structure). With a given oil and choice of the appropriate surfactant and temperature, it is possible to continuously change the ratio of oil to water in a microemulsion from zero to infinity. In other words, it is possible to move in a single equilibrium phase from a water-rich microemulsion containing swollen droplets to an oil-rich microemulsion containing swollen inverted droplets. The structure intermediate between the two extremes is bicontinuous (Fig. 1.12B) and contains sample spanning channels of both oil and water. Again the oil and water are separated by a surfactant sheet, but now the sheet is, on average, not curved toward either the water or the oil.

In a practical sense, the structure of a microemulsion may not be important to a formulator anxious to solubilize, for example, a given oil in water. Of more importance is the nature of the phase diagram, which shows the compositions and relative location of the microemulsion

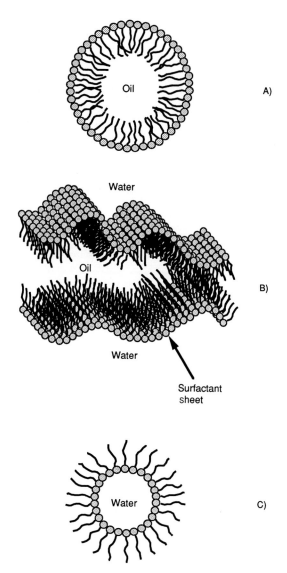

Figure 1.12 Schematics of the structures formed in microemulsions. The microemulsion phase exists below the liquid crystal phase and above the multiphase region. Samples with compositions in the three-phase region contain microemulsions in equilibrium with both oil-rich and water-rich excess phases, and the three-phase region is bounded by two-phase regions (lined): (A) swollen micelle or o/w microemulsion droplet, (B) section from bicontinuous microemulsion showing surfactant sheets separating oil and water, and (C) swollen inverted micelle or w/o microemulsion droplet.

phases. Current thinking is that there probably is not a relationship between phase behavior and structure, but there are guidelines about the kind of structure likely to be formed in a mixture of a given composition. A highly schematic ternary phase diagram (Fig. 1.13) shows the relative location of normal, bicontinuous, and inverted microemulsion structures.

The progression in microstructure and phase behavior is controlled by the nature of the surfactant, the temperature, and the molecular structure of the oil. These parameters can be

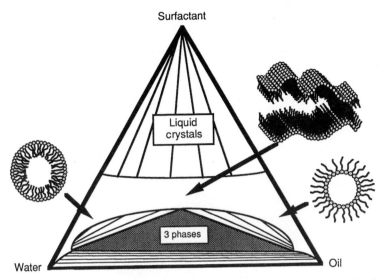

Figure 1.13 Schematic phase diagram for oil, water, and surfactant showing the relative locations of the liquid crystal and microemulsion phases.

further modified or controlled by addition of either "cosurfactants" or salts. Cosurfactants are usually low molecular weight alcohols like butanol or hexanol that modify the surfactant-rich sheet. Such surfactants are thought to make the sheet more flexible and so to promote the formation of small microemulsion droplets, although the actual mode of action is probably considerably more complicated. As discussed above, salts strongly influence the behavior or both ionic and nonionic surfactants and will affect the curvature of the surfactant sheet. Traditional microemulsion formulations found by trial and error often contain at least five components (oil, water, surfactant, alcohol, and salt) and are thus extremely difficult to study in a systematic way. More rational approaches, developed mainly by the group of M. Kahl-weit and R. Strey, have shown that control of temperature and surfactant architecture allows microemulsion formation with the minimal three components: oil, water, and nonionic surfactant. On the other hand, the phase behavior of ionic surfactants is generally not very sensitive to temperature, so microemulsion formation and control of phase behavior for ionic surfactants usually requires the addition of salt or the use of two surfactants.

Space prohibits a full explanation of the factors setting phase behavior and structure, but straightforward application of the ideas already presented can highlight the fundamentals. Consider the evolution of curvature of the surfactant film from, initially, toward oil, to on average flat, to finally toward water. This curvature sets the microemulsion structure, and it can be controlled in several ways. As a practical example, consider the effect of mixing a single-tailed ionic surfactant with a double-tailed ionic surfactant at constant temperature. The surfactants are the single-tailed surfactant dodecyltrimethylammonium bromide (DTAB) and its double-tailed analog didodecyldimethylammonium bromide (DDAB), and the oil is methyl methacrylate (MMA). Figure 1.14A is a sketch of the experimentally determined phase diagram for DTAB/water/MMA at 60 °C. The MMA swells the cores of the DTAB micelles, and the microemulsion region exists near the DTAB/water side of the diagram. In other regions of the diagram, two liquid phases form (2ϕ). In contrast, MMA and DDAB form an inverted structure and the microemulsion region hugs the DDAB/MMA side of the

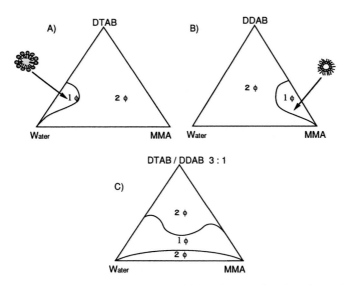

Figure 1.14 Ternary phase diagrams measured at 60 °C for mixtures of water and methyl methacrylate (MMA) with the single-tailed surfactant dodecyltrimethylammonium bromide (DTAB) (A), its double-tailed analog didodecyldimethylammonium bromide (DDAB) (B), and a 3:1 ratio of DTAB/DDAB (C).

diagram (Fig. 1.14B). By blending the single-tailed and double-tailed surfactants, it is possible to open a channel connecting the two sides of the diagram (Fig. 1.14C). In particular, with a surfactant blend of three parts DTAB and one part DDAB, equal amounts of water and MMA can be microemulsified with about 10% surfactant. The microstructure of such a microemulsion is in all likelihood bicontinuous.

The general evolution of phase behavior can be seen best in a phase prism constructed by stacking the ternary phase diagrams of Fig. 1.14 in order of increasing DDAB concentration (Fig. 1.15). In this prism, the weight fraction of DDAB in the surfactant blend is defined as $\varepsilon = DDAB/(DTAB + DDAB)$, and the microemulsion phase rises and moves to the left as ε increases.

This example shows how microstructure and phase behavior can be tuned by adjusting the properties of surfactant molecules. Nonionic surfactants offer a particularly convenient way to adjust phase behavior and "dial-in" structure because their solubility depends so strongly on temperature. The prism analogous to the one shown in Fig. 1.15 can be made with oil and water using a single nonionic surfactant and replacing ε by temperature T. Again, in a practical example, equal amounts of water and tetradecane form a single-phase microemulsion with 14 wt % $C_{12}E_5$ between roughly 40 and 50 °C. At lower temperatures, $C_{12}E_5$ and water solubilizes oil (and the phase diagram is like Fig. 1.14A), while at higher temperatures $C_{12}E_5$ and tetradecane solubilize water (ca. Fig. 1.14B).

Both the temperature range of microemulsion formation and the necessary concentration of surfactant can be adjusted by changing the type or kinds of surfactant used. We now know a tremendous amount about the rational design of microemulsion formulations (see Section 1.8).

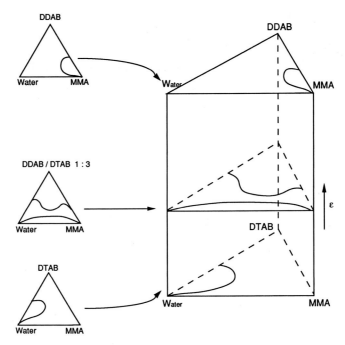

Figure 1.15 The progression of phase behavior in mixtures of MMA/water/DTAB/DDAB as a function of the fraction of DDAB in the surfactant blend. $\varepsilon = $ DDAB/(DDAB + DTAB).

1.6   Surfactant Properties at Interfaces

Having catalogued the behavior of surfactants in bulk solutions, we now turn to a description of the action of surfactants at interfaces. The kinds of microstructure present in surfactant solutions are critical in setting the bulk phase properties like viscosity and are of course central to the ability of an aqueous solution to solubilize oil. On the other hand, interfacial properties are also important in determining the behavior of bulk phases, and most obviously control the wetting or spreading of one material on another.

As a starting point, the difference between a wetting and nonwetting liquid is shown in Fig. 1.16. In Fig. 1.16A, the drop of liquid B does not wet the solid–vapor (S/V) interface. The angle θ between the solid–liquid (S/L) and liquid–vapor (L/V) interface is called the contact angle. When phase L wets the S/V interface (Fig. 1.16B), the contact angle is zero. The contact angle is discussed further below, but for now it is sufficient to realize that the angle can be measured with, for example, a microscope and goniometer. Control of the wetting (and related adhesion) properties of materials is crucial for the successful formulation of paints, dyes, and coatings, and in the application of pharmaceutical and agricultural products.

The most important energy term describing the behavior of interfaces is the interfacial tension, γ, which, under conditions of constant temperature and pressure, represents the work per unit area necessary to create an interface. (Deformations of an interface at constant area are described by bending and twisting moduli, but they are not of importance for this discussion.) From this thermodynamic point of view, the process of wetting will

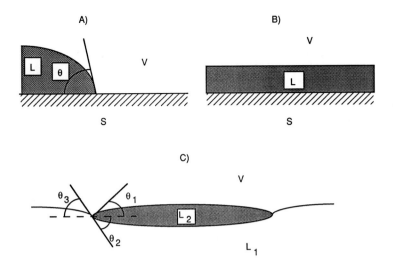

Figure 1.16 Various wetting states for liquid on a solid or a liquid: (A) nonwetting liquid making a contact angle θ with a solid, (B) complete wetting, and (C) nonwetting of an oil droplet at an air–water interface.

occur spontaneously only if the energy of the wetted state (Fig. 1.16B) is lower than that of the dry one. In other words, per unit area it must be favorable to trade the energy $\gamma_{S/V}$ for the energy represented by the creation of the solid–liquid and liquid–vapor interfaces; that is,

$$\gamma_{S/V} > \gamma_{S/L} + \gamma_{L/V} \text{ (wetting condition)} \tag{1.4}$$

The interfacial tension also has a mechanical interpretation as the force per unit length on the edges of an interfacial region. If phase L does not spread on the S/V interface, the condition of mechanical equilibrium requires that the forces active at the three-phase contact balance. The horizontal component of this balance (Fig. 1.16A) is called Young's equation:

$$\gamma_{S/V} = \gamma_{S/L} + \gamma_{L/V} \cos \theta \tag{1.5}$$

Since θ is greater than zero if the liquid does not wet the solid, cos θ is less than 1 for any nonwetting situation. Thus any nonwetting geometry requires

$$\gamma_{S/V} < \gamma_{S/L} + \gamma_{L/V} \text{ (nonwetting condition)} \tag{1.6}$$

which is consistent with the criterion just arrived at by thermodynamic considerations.

The inequalities of the relative combined values of the interfacial tensions in the expressions above suggest the definition of the spreading coefficient:

$$S_{L/S} = \gamma_{S/V} - (\gamma_{S/L} + \gamma_{L/V}) = \gamma_{S/V} - \gamma_{S/L} - \gamma_{L/V} \tag{1.7}$$

The liquid will spread on the solid–vapor interface if $S_{L/S}$ is positive, while the liquid will sit in a drop on the surface if $S_{L/S}$ is negative. Thus, calculation of the spreading coefficient apparently offers an easy way to determine whether a liquid will wet a solid. It is unfortunate that evaluation of the various interfacial tensions required, particularly those involving the solid, is not always straightforward. Liquid–vapor interfacial tensions are tabulated and can be measured in fairly routine ways using, for example, commercially available instruments employing the DuNouy ring or the Wilhelmy plate method. On the other hand, the solid–vapor and solid–liquid tensions cannot be measured in such a direct way.

The first complication is that $\gamma_{S/V}$ is just that—the interfacial energy of the solid in the presence of the vapor from the liquid and other gases in the system. It is *not* the surface energy of the bare solid surface in vacuum, which is called γ_S^0. Bare solid surfaces usually adsorb other molecules, so the value of $\gamma_{S/V}$ relevant to Young's equation will be less than the surface energy γ_S^0. The difference between the two is defined as Π_e, the absorbate spreading pressure (a two-dimensional pressure with units of force per length), via

$$\gamma_{S/V} = \gamma_S^0 - \Pi_e \tag{1.8}$$

It is difficult to measure Π_e, and this lack of knowledge makes it impossible to calculate the contact angle with Young's equation. In practice, Π_e is often neglected, which can lead to substantial errors if the surface has a high affinity for adsorption. Examples of such high energy surfaces are minerals, glass, and most metals. On the other hand, there is a wide class of materials with low surface energies, and these materials do not adsorb the vapor of most liquids. Most polymer surfaces are of low energy, with polytetrafluorethylene (Teflon) being the most famous. For these surfaces, $\Pi_e = 0$.

The spreading of one material on another can be changed by the addition of surfactants. Surfactants generally act to lower the interfacial tension of liquids $\gamma_{L/V}$, and a glance at the definition of $S_{L/S}$ shows that addition of a surfactant may in general change $S_{L/S}$ from negative to positive if the other interfacial tensions are unchanged by the surfactant. Of course, a surfactant will also modify the surface–vapor and surface–liquid tensions, but in general the addition of an appropriate surfactant will promote wetting. The wetting of porous materials and fibers is an important practical problem in the coatings and textile industries. The wetting of the pores of such materials is dramatically aided by the addition of surfactant.

The contact angle geometry as sketched in Fig. 1.16 is in several ways an idealization, and many complications with real liquids on real solids can make the reproducible determination of the contact angle difficult. Indeed it is observed experimentally that the contact angle for a material moving over a surface (the advancing contact angle) is different from that of the same material retreating from an interface (the receding contact angle). The receding angle is usually substantially lower than the advancing angle. The difference comes from several sources. First, most surfaces are not clean, and the presence of contamination will cause the advancing angle to be higher than the contact angle made by the liquid as it withdraws over itself. Surface roughness also contributes to contact angle hysteresis, as do the details of molecular organization at the liquid–solid or solid–vapor interfaces. Evidently contact angle hysteresis is not observed in the spreading of highly pure liquids on rigorously clean surfaces, but it is an important practical problem.

The next complication is due to the vertical component of the force balance at the vapor–liquid–solid contact line. This vertical tension is balanced by stresses in the solid, and these stresses can be high enough to cause deformation of the surface, or at least to alter the local arrangement of surface atoms and thus change the solid surface energy. As a consequence, a liquid drop on a solid with a finite contact angle is in a stable but nonequilibrium configuration.

As the interface supporting a lens of fluid becomes deformable, the vertical component of the force balance can also be satisfied. This is easily seen in the contact of a lens of liquid at the interface of two other immiscible liquids (Fig. 1.16C). For this situation, Young's equation is

$$\gamma_{L1/V} \cos \theta_3 = \gamma_{L2/V} \cos \theta_1 + \gamma_{L1/L2} \cos \theta_2 \tag{1.9}$$

The addition of surfactant in this case may also promote spreading by changing the interfacial tensions, but could also promote the formation of microemulsions or emulsions.

1.7   Emulsions

Emulsions are one of the most useful tools of the formulation chemist. Emulsions are dispersions of one liquid phase within another, usually with the aid of a surfactant. They are thermodynamically unstable, and much of the huge body of empirical or semiempirical work on emulsions has been aimed at improving their kinetic stability or "shelf life." Emulsions are usually either oil-continuous and contain water droplets (so-called water-in-oil) or water-continuous and contain oil droplets (oil-in-water). The physical picture of such structures is much like the sketch of the corresponding microemulsion droplets (Fig. 1.12), but (macro)emulsion droplets are much larger, with diameters ranging from 0.2 to 50 μm. Such emulsion droplets are easily imaged with an optical microscope.

Emulsions are fundamentally different from microemulsions, and the two terms should be used carefully. Emulsions are thermodynamically in a state of two-phase equilibrium, with one of the phases finely dispersed in the other. The dispersion can be made mechanically and kinetically stable by skillful preparation, but the structure is not in equilibrium. Microemulsions on the other hand are single phases, and the microstructure present is thermodynamically stable. Microemulsions are *not* emulsions with small droplets!

The two chief challenges to the formulator are first to make an emulsion of the desired materials, and then to keep it stable. Making an emulsion requires at least the reduction of the interfacial tension between the two liquids. This is most easily accomplished by adding a surfactant, and there are rough guides to predict the action of a particular surfactant in forming an o/w or w/o emulsion. As an aside, we note that emulsions can also be made in the presence of small solid particles that accumulate at the oil–water interface and change the interfacial properties. The stabilizing power of powders depends on the wetting and contact angle properties of the solid and both liquids.

Given the arguments about surfactant packing number in Section 1.3, it is likely that a surfactant that forms normal micelles in water (i.e., one with a low surfactant number) would be effective in forming the surfactant skin around an oil droplet and would thus promote formation of an oil-in-water emulsion. Similarly, a surfactant with a surfactant number near one forms inverted structures in oil and so would promote the formation of a water-in-oil emulsion. Such trends are in fact observed and have led to the general rule of thumb that the phase in which the surfactant is more soluble will be the external phase of a subsequent emulsion. Of course "soluble" in this sense means that micellar structures may be present in the external phase.

The solubility of a surfactant in water at a fixed temperature depends ultimately on surfactant structure and is often reflected in a numerical value called the hydrophile–lipophile balance, or HLB number. The HLB concept contains the same kind of information as the surfactant number, but there is a substantially larger amount of empirical data about HLBs. HLBs range from 0 to 40; high HLB numbers mean that the surfactant is soluble in water. Thus surfactants with high HLBs form o/w emulsions and low HLB surfactants form w/o emulsions. Table 1.5 gives the HLB scale.

The HLB of a surfactant is determined at a fixed temperature, and, like the surfactant number, the HLB number depends on the molecular structure of the surfactant. An HLB scale has also been constructed for various oils, and the best chance of forming an emulsion occurs when the HLB of the surfactant matches that of the oil. The most economical or efficient surfactant of the many surfactants with the same HLB must still be found by experiment.

Because the HLB number of a surfactant is determined at a fixed temperature, the HLB method does not recognize or account for the dramatic changes in surfactant properties with

Table 1.5   The Hydrophile–Lipophile Balance (HLB) Scale

Surfactant solubility in water	HLB number	Comments
No dispersibility or solubility	0–4	W/o emulsifier; high surfactant number
Poor dispersibility unstable turbid dispersion	4–8	W/o emulsifier; surfactant number 0.5–1
Stable turbid dispersion	8–10	Wetting agent
Translucent to clear solution	10–14	Detergent surfactant number < 0.5
Clear solution	14–18	Solubilizer, o/w emulsifier; low surfactant number

temperature, particularly when the surfactant is nonionic. As discussed above, nonionic surfactants change from water-soluble (high HLB number) at low temperatures to oil-soluble (low HLB number) at high temperatures. There is now much active research aimed at finding the reasons for this progression in properties. In one of the first attempts to account for this phenomenon, the temperature at which an emulsion containing equal amounts of oil and water was observed to change from o/w (low temperature) to w/o (high temperature). This temperature, called the phase inversion temperature or PIT, is closely connected to the related *micro*emulsion phase behavior. Variations in temperature during the processing of emulsions with nonionic surfactants has led to the production of especially stable emulsions.

Once an emulsion has been formed, the only remaining important issue is its stability. An emulsion can fail in two ways. The first is flocculation, in which the emulsion droplets aggregate but do not fuse or lose their identity. The floc can eventually settle or float; in either case the emulsion becomes too heterogeneous for use. Weakly flocculated emulsions can be redispersed by stirring or other agitation. The second mode of failure is coalescence, in which the droplets merge into larger droplets. Eventually the droplets become so large that a second phase separates from the emulsion.

The technology for preventing droplet coalescence is well developed and is discussed at length by M. Rosen, A. Adamson, and P. Becher (see Section 1.8). The basic ideas are intuitive and easy to summarize. Clearly the interfaces of droplets in a stable emulsion should be elastic and should resist droplet fusion. The rate of fusion depends in several ways on the viscosity of the continuous phase and on temperature. Emulsion stability is promoted if some kind of energy barrier exists between emulsion droplets, and such a barrier can be engineered by introducing electrostatic or steric repulsions between the droplets. Electrostatic repulsions are generated by ionic surfactants, and steric repulsions can be adjusted by the addition of suitable polymers. Finally, emulsions in which droplets are uniform in size are more stable than those with broad size distributions, because large drops grow at the expense of small ones by Ostwald ripening. Extremely stable emulsions have been made by carefully fractionating the droplets to give narrow size distributions.

1.8   Bibliographic Notes

There are several excellent textbooks in the area of surfactant and colloid chemistry. The most modern encyclopedic text is the two-volume set edited by R.J. Hunter, *Foundations of Colloid Science* [1], which is also available in paperback. A.W. Adamson's *Physical Chemistry of Surfaces* [2] is a classic text, and there is much valuable practical information in *Surfactants and Interfacial Phenomena,* by M.J. Rosen [3].

The ideas of surfactant packaging began with C. Tanford in *The Hydrophobic Effect* [4] and were expanded by J. Israelachvili in *Intermolecular and Surface Forces* [5]. More information about the L_3 phase is given in papers by Porte et al. [6] and Strey et al. [7].

Cmc data are available in a U.S. Department of Commerce publication, *Critical Micelle Concentrations of Aqueous Surfactant Systems* [8.]. The cmc and cloud point data cited in this chapter are from work by Haak et al. [9], Kresheck [10], and Rupert et al. [11].

Modern developments in microemulsion phase behavior and structure are given in many papers by M. Kahlweit and co-workers. References [12] and [13] are useful places to start. Emulsion technology, which is much more developed, is discussed in the texts cited above [1–5], as well as in the *Encyclopedia of Emulsion Technology* [14], edited by P. Becher, and *Emulsions and Emulsion Technology* [15], edited by K.J. Lissant.

Acknowledgments

It is a pleasure to acknowledge the useful assistance of K.A. Marritt in preparing this chapter, as well as the invaluable artistic help of K.F. Kaler.

References

1. Hunter, A.J., Ed., *Foundations of Colloid Science,* Vols. I and II, Clarendon Press, Oxford, 1987.
2. Adamson, A.W., *Physical Chemistry of Surfaces,* 4th ed., Wiley, New York, 1982.
3. Rosen, M. J., *Surfactants and Interfacial Phenomena,* Wiley, New York, 1978.
4. Tanford, C., *The Hydrophobic Effect,* Wiley, New York, 1980.
5. Israelachvili, J., *Intermolecular and Surface Forces,* 2nd ed., Academic Press, Orlando, FL, 1992.
6. Porte, G., Marignan, J., Bassereau, P., May, R., J. Phys. Fr., *49,* 511 (1988).
7. Strey, R., Winkler, J., Magid, L., J. Phys. Chem., *95,* 7502 (1991).
8. *Critical Micelle Concentrations of Aqueous Surfactant Systems,* U. S. Department of Commerce, Washington, DC, 1971; publication NSRDS-NBS 36.
9. Haak, J.R., van Os, N.M., Rupert, L.A.M., *Physico-Chemical Properties of Surfactants: II. Cationic Surfactants,* AMER.88.020, Koninklijke/Shell Laboratorium, Amsterdam, 1989.
10. Kresheck, G.C., in *Water—A Comprehensive Treatise,* F. Franks, Ed., Plenum Press, New York, 1975.
11. Rupert, L.A.M., Haak, J.R., van Os, N.M., *Physico-Chemical Properties of Surfactants: III. Nonionic Surfactants,* AMER.88.021, Koninklijke/Shell-Laboratorium, Amsterdam, 1989.
12. Kahlweit, M., Strey, R., Haase, D., J. Phys. Chem., *89,* 163 (1985).
13. Kahlweit, M., Strey, R., Firman, P., Hasse, D., Jen, J., Schömacker, R., Langmuir, *4,* 499 (1988).
14. Becher, P., Ed., *Encyclopedia of Emulsion Technology,* Dekker, New York, 1983.
15. Lissant, K. J., Ed., *Emulsions and Emulsion Technology,* Dekker, New York, 1974.

CHAPTER 2

Mechanisms of Soil Removal

Guy Broze

Soil removal results from complex interactions between the soil, the substrate, and the cleaning solution. Thus cleaning compositions must be adapted not only to the type of soil but also to the substrate. Substrates found in the household can be hard (e.g., glass, ceramic, metal, plastic) or fabric (e.g., cotton, polyester, polyamide) surfaces.

The main types of soils are:

- Water soluble (easy to remove with clean water but a challenge in dry cleaning!)
- Oily (mainly hydrocarbons, no polar group)
- Greasy (essentially fatty materials containing polar groups such as esters and carboxylates)
- Particulate (e.g., metal oxide, alumina, silica, clay)
- Bleachable (oxidizable soils such as stains from fruit, wine, coffee, and tea)
- Protein and starch (e.g., blood, milk, grass)

Three main mechanisms can contribute to the removal of oily–greasy soils:

- Roll-up, for liquid soils only, is based on the reduction of substrate–water and soil–water interfacial tensions.
- Emulsification is related to the adsorption capacity of the surfactant system. Optimum efficacy is obtained in the neighborhood of the critical micelle concentration of the surfactant system.
- Solubilization is related to the formation of surfactant micelles swollen with oil or grease. This mechanism is found mainly with nonionic surfactants at concentrations above the critical micelle concentration.

The particulate soils are wetted by surfactants and dispersed by mechanical action. Bleachable soils are oxidized by hypochlorite or oxygen bleaches; protein and starchy soils are best digested by adequate enzymes.

Redeposition can be prevented or delayed thanks to electrostatic or, better, steric repulsions between the particles.

In hard water areas, fabric encrustation can be controlled by alkaline earth sequestering agents such as sodium tripolyphosphate, citrate, or zeolite, to which can be added a scale inhibitor, usually sodium polyacrylate.

2.1   Introduction

The general challenge of household cleaning products is to remove completely and rapidly a great variety of soils from a great variety of substrates with products that are safe for the

Guy Broze, Colgate-Palmolive, Research & Development Inc., Avenue du Parc Industriel, B-4041 Milmort (Herstal), Belgium

user, for the surface to be cleaned, and for the environment. Cleaning is a complex task, depending not only on the formula composition but on the nature of the substrate, the soil, and the way cleaning is conducted: product concentration, water hardness, and temperature are among the important variables here. The formulator can determine the composition of the product and make recommendations on how to use it, but he or she does not control the other parameters. To formulate effective products, it is a good idea to develop a system whose performance is as independent as possible of uncontrolled factors.

To meet this challenge, the formulator needs to acquire an excellent understanding of the needs and practices of the consumers, and this includes knowing under what conditions people will use the product. With this knowledge, the formulator will usually try to identify the components of the formulation to meet consumer expectations. This implies a thorough knowledge of the nature of the soils to be removed.

This chapter reviews the main types of household soils, then focuses on the major mechanisms of soil removal, and finally addresses problems related to soil redeposition and fabric encrustation.

2.2   Types of Soils

2.2.1   Water-Soluble Soils

Many soils are soluble in water. Typical examples are salts, such as sodium chloride left after the evaporation of cooking water, and sugar—resulting, for instance, from the spillage of a soft drink. Such soils are best removed with clean, cold water and are seldom of great concern for the housekeeper. They may be an issue, however, when they occur on items that must be dry-cleaned. This is because dry cleaning is based on organic solvent extraction, and water-soluble materials usually are not soluble in the organic solvents used in this process.

2.2.2   Oily and Greasy Soils

By *oily soils* we mean soils mainly composed of nonpolar hydrocarbons such as diesel and motor oils. These materials are usually liquid and highly hydrophobic, which means that they do not mix with water. The interfacial tension between oil and water can be as high as 50 mN/m, which means that forcing contact between oil and water requires a great deal of energy. Because their surface energy is low (20–25 mN/m), the nonpolar hydrocarbons usually spread easily on the substrate. These oils are very soluble in many organic solvents and particularly in halogenated hydrocarbons. They are quite easily removed by dry cleaning. Dry cleaning is not always possible, however, for if the oil is spilled on a floor or on wall paper, it is not advisable to use solvents.

An important nonpolar soil is squalene, a high molecular weight, unsaturated hydrocarbon present in the skin. Since the function of squalene is to protect the skin, ideally it should not be removed from the skin. Skin irritation (erythema) can result if this precaution is neglected.

Greasy soils mainly refer to triglycerides and their derivatives: mono- and diglycerides, fatty acids, and so on. These materials are either animal or vegetable in origin. They are more polar than the oily materials but not polar enough to be dissolved by water. Their physical

Table 2.1   Increase in Triglyceride Chain Length
as a Function of Melting Temperature

Triglyceride	Chain length	Melting temperature (°C)
Trilaurin	12	46
Trimyristin	14	55
Tripalmitin	16	64

state depends mainly on the length and degree of unsaturation of the fatty chain. Fully saturated triglycerides with at least 12 carbon fatty chains are semicrystalline at room temperature. The maximum melting temperature of a pure saturated triglyceride increases with chain length, as illustrated in Table 2.1.

Greasy soil must be regarded as a suspension of small crystalline particles in an amorphous, high viscosity matrix. For a given triglyceride, the melting temperature increases if the crystalline particles are bigger (crystals better formed). The temperature required to melt the grease obviously increases with the proportion of crystalline material. In general, the higher the crystallinity of a grease, the more difficult is its removal.

If a grease contains at least one carbon–carbon double bond per fatty chain (triolein), its melting point is reduced below room temperature. Most unsaturated triglycerides (e.g., peanut, soya, or corn oil) are liquids and accordingly should be easier to remove. This is unfortunately not the case, however, because the presence of the unsaturation is responsible for chemical reactions induced mainly by temperature. When unsaturated greases are heated (cooking), the double bonds can be oxidized and polymerization takes place, leading to an insoluble mass.

2.2.3   Particulate Soils

Solid particles such as clay, alumina, silica, iron, and other metal oxides are present in particulate soils, deposited mostly from air suspensions (dust). They are soluble neither in water nor in organic solvents. They usually exhibit a large surface area, on which the oils and greases absorb very strongly. Particulate soils contribute significantly to the difficulty of removing oily and greasy soils because they contribute to their rigidification and, sometimes, they act as catalysts in the oxidation/crosslinking of unsaturated triglycerides.

Since they are not water soluble, the particulate soils can be redeposited on surfaces that have been cleaned. It is accordingly important to keep such soils effectively dispersed in the washing liquid.

It is noteworthy that particulate soils are not necessarily mineral. Some biopolymers such as starch remain insoluble in water during the dishwashing operation and behave actually as particulate soil.

2.2.4   Oxidizable or Bleachable Soils

Some soils (wine, blood, fruit juices, grass, tea, coffee, etc.) are highly colored, the color resulting either from a series of conjugated double bonds or from porphyrinic structures (wine). These soils can be oxidized by hypochlorite, hydrogen peroxide, or peracids, leading in most cases to colorless substances. The oxidized substances may not be removed by the cleaning operation, but they are no longer visible. It must also be pointed out that even if the soil is not completely removed during the cleaning operation, it is usually broken into smaller pieces, which can be removed in the next wash.

2.2.5   Proteins and Starchy Soils

Proteins and starch are polymeric materials that can resist conventional cleaning. They act as glue for other soils, making cleaning more difficult. A typical example is macaroni and cheese. These soils are best addressed by enzymatic cleaning. Proteolytic and amylolytic enzymes are currently used for this purpose in modern automatic laundry detergents.

2.3   Major Mechanisms of Soil Removal

We now turn to the main mechanisms of soil removal. Water-soluble soils are best removed with clean water by a simple solubilization mechanism. We will not elaborate on this type of soil. Rather, we focus directly on a more difficult task, the removal of oily–greasy soils.

2.3.1   Oily–Greasy Soils

Oily and greasy soils are not water soluble and cannot be satisfactorily removed with pure water. As a consequence, there is a tension at the interface between the soil and water. This tension, which corresponds to the energy needed to bring the oil in contact with the substrate, arises because the cohesive energy of the oil is not the same as that of the substrate. Water is not able to wet the surface of the soil.

Surfactants need to be added to water to reduce the interfacial tension to a reasonable value, allowing water to wet the soil, which actually means the displacement of air by the aqueous medium. This stage is essential because it must be completed before any detergency can take place. Depending on the nature of the soil, the surfactant used, the temperature, and the mechanical action, different things can happen [1–3].

2.3.1.1   Roll-Up

If the soil is liquid, either an oil or a grease above its melting temperature, the so-called roll-up mechanism can take place [4]. When a liquid oily soil O lies on the surface of a solid substrate S, there is an interfacial tension γ_{OS}, at the soil–substrate interface (Fig. 2.1).

The oil droplet forms with the substrate an angle θ, which results from the balance between the surface forces. The surface tension γ_{SG} between the substrate and the gas phase (air in the present case) tends to spread the oil on the surface (to reduce the surface of contact). The interfacial tension γ_{OS} between the oil and the substrate tends to reduce the oil–substrate contact and to re-form the drop. The surface tension γ_{OG} between the oil and the gas phase is

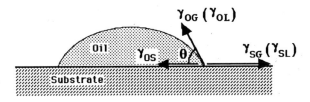

Figure 2.1   Drop of oil on a solid substrate.

not in the plane of the substrate. Its projection on the plane nevertheless must be taken into account to balance the forces. The equilibrium condition is given by eq 2.1, known as Young's equation.

$$\gamma_{SG} = \gamma_{OS} + \gamma_{OG} \cos \theta \qquad (2.1)$$

The interfacial tension between the oil and the substrate is usually lower than the surface tension between the substrate and the gas phase, especially on organic surfaces such as fabrics and polymeric hard surfaces; the oil surface tension seldom exceeds 25 mN/m. The result is that the contact angle is usually small, which is synonymous with good wetting, and sometimes zero, which is synonymous with perfect wetting.

When the substrate and the oil are submerged, the surface tension between the substrate and the gas phase is replaced by the interfacial tension γ_{SL} between the substrate and water, and the surface tension between oil and gas is replaced by the interfacial tension γ_{OL} between the oil and water. The result is usually another contact angle θ'.

If an appropriate surfactant is dissolved in water, the effect is a strong reduction of the substrate–water and oil–water interfacial tensions, due to the adsorption of surfactant molecules at the water–oil and water–substrate interfaces. Of course, the oil–substrate interfacial tension is not affected, since water does not diffuse to that interface. The system evolves toward a new equilibrium state, characterized by a higher contact angle. Mechanical action causes droplets of oil to be removed from the substrate, and only a small quantity of oil is left (Fig. 2.2).

It is evident that the higher the equilibrium contact angle, the more easily the soil is removed. If the reduction of interfacial tension is so strong that the sum of the oil–water and substrate–water interfacial tensions reaches the oil–substrate interfacial tension, the contact angle is 180° ($\cos \theta = 1$), which means that no oil is left on the substrate (spontaneous perfect cleaning).

It must be pointed out that the cleaning operation depends significantly on the kinetics of the operation. The surfactant needs to diffuse rapidly to the surfaces to reduce the interfacial forces. Hydrodynamics also plays a key role, inasmuch as roll-up implies the motions of fluids. Cleaning based on roll-up is more difficult on irregular substrates such as

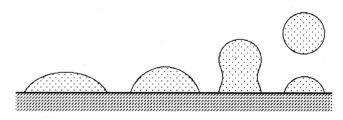

Figure 2.2   Roll-up mechanism.

fabrics; capillary forces are almost unpredictable in real cases, and the result is often incomplete soil removal.

Roll-up has been visualized in real-life conditions by high speed cinematography [5].

2.3.1.2   Emulsification

Emulsification is probably one of the most important mechanisms contributing to oily–greasy soil removal. An *emulsion* is a dispersion of a liquid in another liquid in which it is not soluble. Milk is a typical emulsion. The dispersed droplets usually are not bigger than a few micrometers. Mechanical agitation is able to generate an emulsion, but, to remain dispersed, the oil droplets need to be coated with a layer of surfactant molecules to reduce the interfacial tension with the continuous phase. Since the surface area increases when the droplet size decreases, efficient surfactant systems are required to achieve small particle sizes.

When surfactant is added to pure water, the interfacial tension between oil and water drops, as illustrated in Fig. 2.3. This reduction occurs because the surfactant molecules adsorb at the surface of the oil and reduce the number of contacts between oil and water. Since a surfactant contains a part that is completely water soluble (its polar head group), the interfacial tension could go to very low values and even completely vanish. Unfortunately, such an extreme situation never occurs, owing to surfactant shape and micellization.

To maximize its adsorption, a surfactant must establish complete coverage of the surface, to prevent oil–water contacts. On the molecular scale, the surface of an oil droplet, even as small as 1 μm, is almost planar (like the surface of the earth seen by ground-based humans). Perfect coverage of the surface requires the hydrodynamic (solvated) volume of the hydrophilic part of the surfactant to be equal to the hydrodynamic volume of its hydrophobic part. Under such conditions, the interfacial tension can reach values below 0.1 mN/m. If this is not the case, the coverage of the oil–water interface is imperfect and higher interfacial tensions result.

There is a thermodynamic equilibrium between the surfactant concentration in water and the surfactant adsorbed on any surface, and particularly the surface of the oil droplets. Increasing the surfactant concentration in water also increases its adsorption. The formation of surfactant micelles above a critical concentration, the critical micelle concentration (cmc), limits the effective surfactant monomer concentration from increasing further, so the amount adsorbed remains basically constant. Micellization can occur before the interface is totally covered.

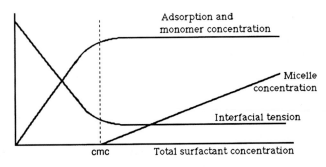

Figure 2.3   Effect of surfactant concentration on adsorption, interfacial tension, and surfactant aggregation (schematic).

Of course, a high surfactant concentration may at first appear to be beneficial, since micelles act as surfactant reservoirs. This is only partly true, however. At surfactant concentrations above 5–10 times the cmc, the micelles can induce flocculation of the oil droplets, which has an adverse effect on the physical stability of the emulsion [6].

The physical stability of the emulsion is an important factor in detergency, but it is not the only one. In fact, in the presence of an appropriate mechanical action, systems with interfacial tensions as high as 5 mN/m can be dispersed, but the maximum cleaning that can be achieved by interfacial tension is also determined. If the interfacial tension is too high, it may no longer be able to balance the oil–substrate interfacial tension. As a result, the equilibrium contact angle of the oil on the substrate is no longer 180° and residual oil is left on the substrate.

2.3.1.3 Solubilization of Oil

Solubilization occurs to a significant extent only in some cases and always above the cmc of the surfactant system. "Solubilization" is not really the appropriate term, inasmuch as oil is not molecularly dissolved in the water phase but is absorbed inside the core of the surfactant micelles.

The solubilization, or oil uptake capacity, of a surfactant system depends on the shape of the micelles. Globular micelles, like those obtained with such highly water-soluble surfactants as anionics (sodium lauryl sulfate, sodium alkylbenzenesulfonate, etc.) exhibit a low oil uptake capacity. Accordingly, solubilization does not contribute significantly to detergency by cleaners based on anionic surfactants. Nonionic surfactants such as polyethoxylated fatty alcohols exhibit a much higher potential for solubilization.

The oil uptake capacity of globular micelles is limited because the addition of oil necessarily results in an increase of the micelle surface exposed to water. Even if the interfacial tension is low, an increase of surface implies an increase of energy that can hardly be balanced by the entropy of dispersion of the oil in the micelles. Moreover, the interfacial tension between a flat oil surface and the aqueous surfactant solution creates a kinetic barrier to spontaneous solubilization, as the adsorption effectiveness is not maximum.

The adsorption effectiveness, which is maximum when the interfacial tension is minimum, can be related to the shape of the surfactant micelles. As shown by Oetter and Hoffmann [7], if the interfacial tension with decane is above 1 mN/m, between 1 and 0.1 mN/m, or below 0.1 mN/m, micelles are respectively globular, rodlike, or disklike.

Rodlike micelles are much better adapted to higher oil uptake. A surfactant forming rodlike micelles induces a lower oil–water interfacial tension and accordingly facilitates the transfer of oil from the substrate to the core of the micelles. The absorption of oil allows micellar internal tensions to relax, and the volume increases to a significant extent without increasing the surface exposed to the water phase.

Rodlike micelles are obtained if the hydrodynamic volume of the hydrophilic group is slightly higher than that of the hydrophobic group. The packing induces a moderate curvature. If the natural radius of curvature of the surfactant molecules exceeds their extended hydrophobic chain length, the resulting hypothetical globular micelles will exhibit a hole in the center and, accordingly, will collapse into a rodlike structure. In rodlike micelles, the hydrophilic heads are compressed in the cylindrical part of the micelles, and the hydrophobic tails are compressed in the hemispherical ends. Such a micelle can absorb oil and restore its natural curvature without any surface area increase.

Rodlike micelles can be obtained with ethoxylated nonionic surfactants in a temperature range just below their cloud point. This is the reason for their good reputation as oil solubilizers. Rodlike micelles can also be obtained with some anionic surfactants in the presence of magnesium or calcium counterions.

If the hydrodynamic volume of the hydrophilic part is almost equal to that of the hydrophobic part, the oil surface coverage is the most efficient, and disk-like micelles are obtained. Such micelles are not very stable. They usually assemble to form a lamellar liquid crystal, which is a very efficient grease cleaner but cannot necessarily maintain the grease in suspension.

From a general point of view, the solubilization capacity of a given surfactant is maximum if it divides equally well between the water and the oily phase. In practice, this happens for a given surfactant system at a temperature referred to as the *phase inversion temperature*. The phase inversion temperature depends on the surfactant structure, on the oil, and on the presence of ingredients dissolved in water. Kahlweit and co-workers gave an extensive view on how the phase inversion temperature varies with the parameters of an oil–water–surfactant system [8–10].

2.3.2   Particulate Soils

Particulate soils almost always occur with other soils such as oily and greasy soils. The particulate soils contribute to the toughness of the soil deposit, and the grease acts as a cement, binding the particles together. The first step, just after wetting, is to attack the oily–greasy component. The particulate soils are then made available.

The best way to clean particulate soils is to use a surfactant that is able to adsorb efficiently at the water–solid particle interface, to reduce the interfacial tension and, accordingly, to reduce the adhesion forces binding the particles together. This can be achieved with an anionic surfactant, in which case the surface of the solid particles is made more negative and electrostatic repulsion can occur between adjacent particles.

Since particulate soils are not water soluble, they have a tendency to redeposit in the later stages of the washing operation. The problem of the soil redeposition is treated in Section 2.4.1.

2.3.3   Bleachable Soils

In laundry, bleachable soils are treated with an oxidant. The oxidant can be based on either hypochlorite or oxygen. Sodium hypochlorite, the most common bleaching agent, is widely used in North America as a laundry additive. It is inexpensive, very efficient, and delivers a sanitation benefit. Unfortunately, hypochlorite can damage colors and the fabric itself. The formulation of finished cleaning products with sodium hypochlorite is very difficult because the hypochlorite reacts very rapidly with many organic materials.

Oxygen bleaches have been preferred in Europe and are now being adopted worldwide. The most common system is based on hydrogen peroxide, which is a much milder oxidant than hypochlorite. Sodium perborate tetrahydrate or monohydrate is used as a source of hydrogen peroxide. Since its molecular weight is lower, perborate monohydrate is more concentrated in hydrogen peroxide, but it is a brittle solid. It is difficult to formulate in a powdered detergent because it generates a lot of dust. Perborate tetrahydrate, although less rich in hydrogen peroxide, is more appropriate for the formulation of powders because the beads exhibit good mechanical resistance.

In contact with water, perborate dissolves slowly and releases hydrogen peroxide. The full bleaching capacity of hydrogen peroxide is generated only at 70 °C and above. Effective bleaching can be achieved at lower temperature by using bleach activators [11]. The most common one is tetracetylethylenediamine (TAED).

Peracetic acid is the actual low temperature bleaching agent. It does an effective job even at 40 °C. At higher temperatures, peracetic acid can decompose to regenerate hydrogen peroxide. Moreover, peracetic acid can react with hydrogen peroxide in the presence of colloidal catalysts such as silica. To prevent this undesired reaction, the addition of phosphonate-based stabilizers is recommended.

The activated perborate bleaching system is quite easily formulated in powder detergents. The bleaching system can remain active after several months of aging if the atmosphere is not too humid.

Preformed peracids can also be used as bleaching agents. Magnesium monoperoxyphthalate is an example [12]. Such a molecule is already a peracid and is accordingly active at 40 °C. The peracids are in general harder to stabilize because they are more reactive than activated systems. They can also be responsible for spot bleaching (local discoloration) if solubilization is not fast enough.

Hypochlorite bleaches usually are used in automatic dishwashing, mainly for reasons of sanitization, efficacy in removal of colored stains (e.g., tea, coffee), and ability to break down protein soil.

For more details on bleaching see Chapter 4, and for its application in automatic dishwashing see Chapter 7 (Section 7.2).

2.3.4   Protein and Starchy Soils

Proteins and starches are present in significant quantities in food stains. The moderate water solubility of these polymers makes them difficult to remove with classical techniques. Proteases and amylases are enzymes able to hydrolyze proteins and starch, respectively. Such enzymes are currently used in laundry detergents and even in automatic dishwashing detergents.

The selection of a protease is challenging: since the enzyme itself is a protein, it is subject to self-destruction. To prevent such autohydrolysis, sophisticated stabilizing systems have been developed. An elegant way to proceed to stabilize the enzyme structure in built liquid laundry detergents is to use a borax–glycol complex to reduce the water activity [13].

2.4   Second-Order Challenges

2.4.1   Soil Redeposition

Once the soil has been detached from the substrate, it is either "solubilized" inside the micelles, emulsified, or, in the case of solid particles, dispersed as a suspension in the washing liquid. The emulsions and dispersions of solids are not stable, and often the soil redeposits on the cleaned items or on parts of the washing machine. Redeposition can be assessed by

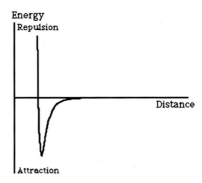

Figure 2.4 Van der Waals interactions: energy profile as a function of the distance between two particles.

measuring the loss of reflectance (whiteness) of white fabrics after several cumulative washes with soiled items. A typical procedure has been proposed by Feighner [14].

The physicochemical phenomenon involved is the following. At short distances, even nonpolar solid particles attract each other, thanks to van der Waals interactions. This type of interaction decreases with the sixth power of the distance between the particles. The energy profile is shown in Fig. 2.4. The repulsion at very short distances is due to penetration of the electronic clouds. This repulsion usually decreases with the twelfth power of the distance (Born repulsion).

The result is an energy minimum at short distances. This minimum is so deep that the particles cannot be separated by mechanical energy. Usually referred to as the *primary min–imum*, this minimum corresponds to irreversible flocculation or *coagulation*.

This attraction is not limited to two particles. A large number of particles can agglomerate to form a big floc that deposits on the washed items. More information on van der Waals forces can be obtained in Hunter [15, p. 168].

There are two basic ways to stabilize a dispersion in an aqueous liquid and to prevent (or delay) redeposition: electrostatic repulsion and steric stabilization.

2.4.1.1   *Electrostatic Repulsion*

If the particles are electrically charged, electrostatic repulsion can create an energy barrier to flocculation/coagulation, represented by ΔE in Fig. 2.5. Electrostatic repulsion drops with the square of the distance. Its magnitude depends on the amount of charge on each particle, the dielectric constant of the solvent, and the ionic strength. If the latter is low, the distance at which the electrostatic repulsion is perceived (Debye length) is relatively long (on the order of 1 μm in water); if the water contains an electrolyte (> 0.01 mole, as NaCl), the Debye length is reduced, as well as the energy barrier ΔE.

It is not a good idea to rely on electrostatic repulsion to prevent redeposition because ionic strength is not a parameter under our control. Stains can include a significant amount of electrolytes, and builders, contained in most detergents, are electrolytes.

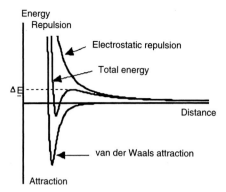

Figure 2.5   Energy profile in the presence of electrostatic repulsion.

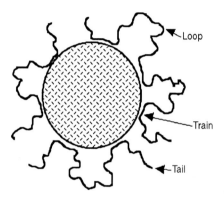

Figure 2.6   Polymer adsorbed on a solid particle.

2.4.1.2   *Steric Stabilization ("Colloid Protection")*

Steric stabilization offers a better potential for the control of redeposition. The principle is to deposit a large molecule (a surfactant or a polymer) on the surface of the particle or of the substrate to be protected from redeposition.

Figure 2.6 illustrates the adsorption of a polymeric material on a solid particle. Some polymer segments called *trains* stick to the surface (thanks to van der Waals or electrostatic interactions), and other segments called *loops* interact with water. The *tails* of the molecule usually interact with water. This process is dynamic: the trains can desorb and the loops can adsorb.

The antiredeposition agent must exhibit the right balance between water solubility and adsorption efficacy. If the water solubility is too high, the proportion of loops is too high and the adsorption is not efficient; on the other hand, if water solubility is too low, there are almost no loops and the molecule is deposited as a thin layer on the surface of the particle.

The result of the expansion of the loops in the water phase is an increase of the effective volume of the particle. If two particles come together, the loops of their adsorbed polymers begin to interact. In the interaction volume, the polymer concentration becomes too high, resulting in an osmotic pressure (water molecules want to enter the interaction volume and pull the particles apart). There is also an entropic effect: compression significantly reduces

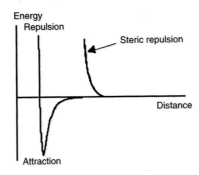

Figure 2.7   Energy profile in the presence of steric repulsion.

the configurational entropy of the polymer. The net result is illustrated in Fig. 2.7. At a distance significantly longer than the onset of van der Waals attraction, a new energy barrier, corresponding to the compression of the polymer loops, develops to prevent flocculation.

The control of redeposition is significantly more complex than this simple theoretical approach would lead one to believe. The adsorption efficacy depends on the nature of the surfaces, on the composition of the water phase, on temperature, and so on. The use of polymer adsorption for suspension stabilization is very delicate. Particularly when too small an amount of a high molecular weight polymer is used, interparticle bridging can occur, speeding up the flocculation.

To prevent redeposition on cotton fabrics, carboxymethylcellulose appears to be the most efficient material. On polyester fibers, hydroxyethyl- or hydroxypropylcellulose gives better results.

For additional readings on the fundamentals of suspensions and their stabilization, refer to Hunter [15, p. 450] and Tadros [16].

2.4.2   Encrustation

Automatic laundering can generate fabric encrustation, particularly in areas where the water is hard. The fabric encrustation can be organic (most often soap deposits) or mineral (lime scale or other insoluble calcium salts). Mineral deposits can be responsible for fabric graying and wear, as well as a surface roughening effect.

The problem of fabric encrustation becomes critical with the reduction of sodium tripolyphosphate in areas where the water hardness reaches or exceeds 300 ppm (as calcium carbonate). In several European countries, for example, the level of encrustation can reach very high levels!

Fabric encrustation is related to the ratio between calcium ions and tripolyphosphate (TPP). If the Ca^{2+}/TPP ratio is not greater than 1, a water-soluble Ca-TPP complex is formed and almost no encrustation is observed. This happens in areas of soft water or, in hard water, if the amount of TPP is high enough. If the Ca^{2+}/TPP ratio exceeds 1, a water-insoluble Ca_2-TPP complex is formed, which can deposit on the fibers. This occurs if water hardness is increased or the level of TPP is reduced.

The quality of the phosphate builder is also essential. Higher encrustation levels can also be observed if TPP is degraded to pyrophosphate and orthophosphate, calcium orthophosphate being water insoluble.

Nonphosphate (no-P) built formulations can be based on sodium citrate (calcium salt soluble) or, more often, zeolite 4A. Zeolite particles can also deposit on fabrics if the suspension is not stabilized properly, but the level rarely exceeds 5%. The problem is more apparent on dark items, on which white zeolite deposits are more apparent.

In "no-P" built products it is a good idea to add several percent of a low molecular weight sodium polyacrylate or acrylate-based copolymer. Such polymers present a moderate complexation power for calcium and magnesium, but they mainly act as scale inhibitors. Rather than sequestering the alkaline earth cation, they reduce the rate of crystallization of their insoluble salts (mainly calcium carbonate) by interfering with crystal growth. Depending on concentration and temperature, sodium polyacrylate polymers can delay calcium carbonate precipitation for several hours. The effect is essentially kinetic, and precipitation will eventually develop.

For more information on polymer applications see Chapter 5.

References

1. Stevenson, D.G., J. Soc. Cosmet. Chem., *XII*, 353, (1961).
2. Schwartz, A.M., J. Am. Oil Chem. Soc., *48*, 566 (1971).
3. Wingrave, J.A., Soap/Cosmet./Chem. Spec., *57*, 33 (Nov. 1981).
4. Rosen, M.J., *Surfactants and Interfacial Phenomena*, 2nd ed. Wiley, New York, 1988, p. 363.
5. Weber, R., Henkel-Referate Int. Ed., *22*, 144 (1986).
6. Bibetter, J., Roux, D., Pouligny, B., J. Phys. II Fr., *2*, 401 (1992).
7. Oetter, G., Hoffmann, H., J. Dispersion Sci. Technol., *9*(5&6), 459 (1988–1989).
8. Kahlweit, M., Strey, R., Angew. Chem. Int. Ed. Engl., *24*, 654 (1985).
9. Kahlweit, M., Strey, R., J. Phys. Chem., *91*, 1553 (1987).
10. Kahlweit, M., Strey, R., Busse, G., J. Phys. Chem., *94*, 3881 (1990).
11. Leigh, A.G., U.S. Patent 4,225,452, 1980.
12. McCrudden, J.E., et al., U.S. Patent 4,154,695, 1979.
13. Boskamp, J.V., U.K. Patent GB-2,079,305, 1980.
14. Feighner, G.C., J. Am. Oil Chem. Soc., *61*(10), 1645 (1984).
15. Hunter, R.J., *Foundations of Colloid Science*, Vol. I, Clarendon Press, Oxford, 1989.
16. Tadros, T.F., *Surfactants*, Academic Press, London, 1984.

CHAPTER 3

Surfactants

Michael F. Cox

More than 15 billion pounds of commercial surfactants are used each year on this planet. The majority of this volume is used in a wide array of products aimed at providing a cleaner and healthier environment for man. The rest is used for various nonhousehold applications including detergents for motor oils, process aids for industrial chemical reactions involving immiscible phases, and cleanup processes for chemical and mining wastes.

This chapter examines what makes a surfactant a "surfactant," discusses what surfactants are used for and why, and tells what types of surfactant are available. In addition, we characterize major commercial surfactants with respect to how they are made, their composition, their performance attributes, and their availability. Finally, we offer suggestions for obtaining additional information.

3.1   Introduction: What Is a Surfactant?

The word *surfactant* is short for "surface active agent." The term is literally descriptive because a surfactant does in fact concentrate at the surface (between immiscible phases), where it affects the surface properties of the interface. Surfactants concentrate at interfaces because surfactant molecules generally consist of two parts (see Fig. 3.1): a water-soluble part (called the hydrophile) and an oil-soluble part (called the hydrophobe or lipophile).

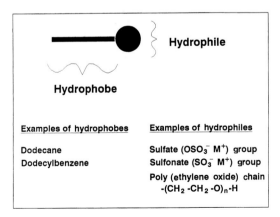

Figure 3.1   Typical surfactant molecule showing the "hydrophile" and "hydrophobe" portions of the molecule.

Michael F. Cox, Vista Chemical Company, P.O. Box 200135, Austin, Texas 78720-0135, U.S.A.

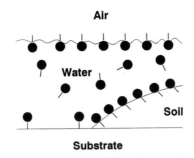

Figure 3.2   Surfactant alignment in an air–water–soil system.

Surfactants concentrate at boundaries between immiscible phases (e.g., air–water interface, soil–solvent interface, fabric–water interface) because an interface allows the surfactant molecule to orient itself such that the hydrophilic portion of the molecule is associated with the more hydrophilic phase of the interface while the hydrophobic portion of the molecule associates with the hydrophobic portion of the interface. For example, if we place a surfactant molecule in water, it will align itself such that the hydrophilic portion of the molecule is associated with water molecules and the hydrophobic portion associates with air or any other surface (walls of vessel, suspended solids, soils, etc.) that provides a more hydrophobic environment (see Fig. 3.2).

The presence of a surfactant at an interface affects the surface properties of the liquid phase. For example, if a surfactant is added to water, it will interfere with the ability of the water molecules to engage in hydrogen bonding with each other at the air–water interface. This results in a reduction in surface tension, which is why a needle floats on pure water but sinks when surfactant is added. Similarly, if surfactant is added to water sitting on oily or waxy soil, the water will spread out over the soil because the interfacial tension between the soil and the water is reduced. This phenomenon is referred to as *wetting*.

Although most of us discuss surfactant performance in terms of a desired end result (foaming, detergency, etc.), basic surfactant properties can tell a great deal about the surfactant molecule itself. Having a basic understanding of surface chemistry will improve your ability to choose the right surfactant for the job. Consequently, readers are urged to study Chapter 1, Basic Surfactant Concepts, and to arrange access to a basic surface science text, such as the one by Rosen [1].

3.2   Basic Surfactant Performance

Surfactants are called on to perform a variety of tasks. Knowing specifically what you want the surfactant to do will significantly help you choose the right surfactant for the job.

A surfactant can help accomplish six basic functions: it can help to wet surfaces (assist in bringing solvent and surface together), facilitate soil removal, suspend materials (solubilize and emulsify), stabilize foam, adsorb on surfaces to amend properties of the surface (e.g., fabric softening), and act as a biocide. No surfactant does all these jobs well, and the ability of any surfactant to accomplish any of these tasks is dependent on its concentration and its environment (temperature, ionic strength, pH, etc.). This is why there is such a variety of surfactants from which to choose.

Although the surface chemistry associated with several of these functions is covered in Chapter 1, we review here the practical aspects of each of these functions, focusing on guidelines one can use to best accomplish each function.

3.2.1   Wetting

Wetting is defined as the displacement of air from a solid or a liquid by an aqueous solution. In more basic terms, wetting refers to the ability of a surfactant solution to spread over a given surface. Surfactants accomplish this by lowering the energy barrier between the solvent and the substrate. The ability to "wet" is a function of several parameters including molecular structure of the surfactant, its concentration, its environment, and the composition of the substrate to be wetted. Consequently, the ability to wet is somewhat difficult to predict.

Wetting can be examined by measuring the "contact angle" of a drop of surfactant solution sitting on the substrate of interest. "Contact angle" is the angle between the substrate surface and the droplet (see Fig. 3.3). The closer the contact angle is to zero, the better the wetting agent (surfactant). One can also calculate a "spreading coefficient" using contact angle and surface tension measurements as represented in the following relationship:

$$\text{spreading coefficient} = \gamma_{LJA}(\cos \theta - 1) \tag{3.1}$$

where γ_{LJA} is the surface tension at the liquid–air interface.

The spreading coefficient, a measure of the energy or driving force behind spreading, is useful in comparing one surfactant with another. Contact angle and surface tension measurements require specialized equipment. If equipment is not available, simple observation of spreading can be useful. Another approach is to compare a droplet diameter of a set volume of solution with that of pure water [2]. Contact angle and spreading coefficient data often can be obtained from surfactant suppliers.

Wetting is important in detergency processes because it helps bring solvent and soil together. However, since other processes (soil suspension/solubilization/emulsification) are

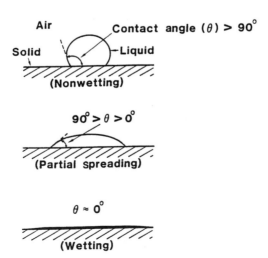

Figure 3.3   Contact angles for nonwetting, partial wetting, and wetting.

also important, formulators often ignore wetting and examine final product performance (detergency performance, foam performance, etc.). With some applications, however, particularly those involving hard-surface cleaning, consideration of wetting can help the formulator. For example, the purpose of a surfactant in a glass cleaner is to wet the glass surface. Since a minimum amount of residue left on the glass is desirable, finding the surfactant that is the most efficient wetting agent (one that works best at the lowest concentration) is desirable.

Two useful guidelines in selecting a surfactant for wetting are as follows:

- Surfactants with intermediate water solubilities are usually better wetting agents than those having low or high water solubility. Differences in water solubility affect wetting properties because they affect the amount of surfactant adsorbed at the solid–liquid interface. If a surfactant is too soluble, it will principally reside in the bulk solution phase. If it is to insoluble, it will not dissolve and will reside in a phase all by itself. Intermediate solubility will allow dissolution and still force the surfactant to concentrate at the interface, where it can lower surface tension and improve wetting. Keep in mind that other ingredients in the formulation, or other materials present during product use, can affect surfactant solubility and therefore affect wetting. Increasing ionic strength can significantly affect the solubility of some surfactants, as can the presence of cations or anions that bind with anionic and cationic surfactants. Do not make the mistake of choosing a surfactant that performs well in pure water but poorly under real use conditions.
- Both anionic and nonionic surfactants should be tested. If the surface of interest is charged, the performance of anionic surfactants can be altered (enhanced or reduced), depending on charge compatibility.

3.2.2   Soil Removal

There are basically three principal mechanisms for removing soil (see Fig. 3.4). Detergency involves the use of surfactants to remove soil through surface chemical processes. In contrast, mechanical processes use physical means (abrasion, etc.), while chemical processes involve the use of solvents or agents (e.g., acids, enzymes, bleaches) that degrade the soil. In most applications, soil removal through detergency is preferred because it offers a more cost-

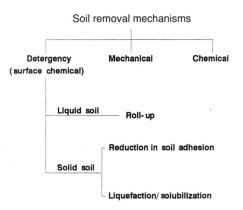

Figure 3.4   Soil removal mechanisms.

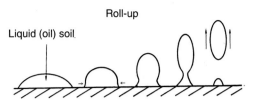

Figure 3.5   "Roll-up" mechanism for removal of liquid oil soils.

effective and versatile approach, which usually does not permanently affect the surface of interest. Most detergency processes, however, also rely on some degree of mechanical action to achieve soil removal (agitation in a washing machine, high pressure sprays, hand rubbing, etc.). Some also rely on chemical solvation of the soil to aid in soil removal (e.g., addition of caustic to saponify and solubilize natural oils and fats).

All detergency processes involve wetting to some degree, but the mechanism varies depending on the soil. For particulate soils, surfactant molecules (and associated water molecules) adsorb onto the surface of both the particulate matter and the surface to be cleaned. This adsorption confers charge of like sign, which helps to lower adhesion because like charges repel. Anionic surfactants are normally considered to be best at removing particulate matter.

Liquid oily soils are usually removed via "roll-up" (see Fig. 3.5). If liquid oily soil is spread over a surface and then submersed in an aqueous solution of surfactant, the surfactant can preferentially wet the substrate surface, leaving less and less surface available for the soil. Once the surface area of the soil has been sufficiently reduced, it can dislodge and float to the surface. Promoting roll-up is best determined empirically, although alcohol ethoxylates are known to excel at removing liquid oily soils via roll-up.

The removal of solid organic soils (waxes, greases, etc.) also involves wetting followed by processes that liquefy the soil. Liquefaction is thought to involve converting the soil, layer by layer, into a more liquid phase (or phases), which aids in mechanical removal. Solid, organic soils are normally important in certain hard-surface cleaning applications. Selecting a surfactant for removing this type of soil is also best accomplished empirically, although alcohol ethoxylates, particularly short-chain ($C_6 - C_{10}$) ethoxylates, have been shown to perform well [3].

3.2.3   Suspend Material (Solubilize/Emulsify)

When soil is removed from surfaces, it is important that it not be allowed to readsorb onto the clean surface (a process called redeposition). Surfactants help accomplish this by adsorbing onto (surrounding) the soil droplets in such a way that the hydrophilic groups are in the aqueous phase and the hydrophobic groups are in the oil phase. This adsorption stabilizes the droplet and helps prevent it from coalescing with other droplets. Although surfactants help in this regard, in most laundry detergents where prevention of redeposition is important, other agents (e.g., carboxymethylcellulose) are added as antiredeposition agents.

It is sometimes necessary to combine aqueous and organic phases to achieve adequate cleaning performance. One example is a tar/asphalt cleaner that requires an aqueous continuous phase for cost and safety reasons and a dispersed organic phase for soil

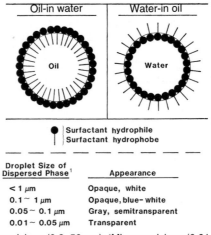

Droplet Size of Dispersed Phase [1]	Appearance
< 1 µm	Opaque, white
0.1 ~ 1 µm	Opaque, blue- white
0.05 ~ 0.1 µm	Gray, semitransparent
0.01 ~ 0.05 µm	Transparent

[1]Macroemulsions (0.2~50 µm); [1]Microemulsions (0.01~0.2 µm).

Figure 3.6   Oil-in-water and water-in-oil emulsion types and typical appearances.

dissolution. Since most organic materials are insoluble in water, the only way to formulate such a product is to make an emulsion.

An *emulsion* is a stable dispersion of one liquid in another (immiscible) liquid. Emulsions require an "emulsifying agent," which normally consists of a surfactant or surfactant mixture. Emulsions are important in the preparation of a variety of products including cosmetic cleaners and creams, and some hard-surface cleaners.

There are two types of emulsions: oil dispersed in water (o/w) and water dispersed in oil (w/o) (see Fig. 3.6). The types vary in appearance, from transparent to milky-white, depending on the size of the droplets making up the dispersed phase [4].

Although formulation of an emulsion can be accomplished empirically, several methods are available to help. One useful tool in selecting a surfactant for an emulsion is the hydrophile–lipophile balance (HLB) method. HLB is defined as the ratio between a molecule's hydrophilic and lipophilic (hydrophobic) parts. From a practical standpoint, it is a method for assigning to emulsifying agents (and the substance to be emulsified) a numerical value (0–40) that helps in predicting their emulsification behavior. HLB numbers are useful because they provide a correlation between materials to be emulsified and agents well suited to emulsify them. When the HLB of the substance to be emulsified is known, a surfactant or surfactant blend is chosen which gives the same HLB number.

The HLB number of a surfactant can be obtained from literature sources (e.g., *McCutcheon's* [5]), determined experimentally [6], or estimated by empirically derived relationships that correlate surfactant structure to HLB behavior [7,8]. For example, the HLB of alcohol ethoxylates can be estimated by dividing the weight percent ethylene oxide (EO) content by 5.

$$\text{HLB (ethoxylates)} \approx \frac{\%EO}{5} \tag{3.2}$$

HLB values can also be averaged. For example, a mixture of 25% surfactant A (HLB = 8) and 75% surfactant B (HLB = 14) gives an HLB of 12.5.

$$(0.25 \times 8) + (0.75 \times 14) \approx 12.5 \tag{3.3}$$

Table 3.1 Commonly Emulsified[1] Materials versus HLB of Surfactants Empirically Determined to Best Emulsify Them

Materials	HLB
Paraffinic mineral oil	10
Paraffin wax	10
Aromatic mineral oil	12
Kerosene	14
Mineral spirits	14
Lauryl alcohol	14
Pine oil	16
Stearic acid	17

[1] Oil-in-water emulsions.

Table 3.2 Example of How Surfactant Blends are Used to Obtain Most Stable Emulsion

Surfactant/surfactant blend	HLB
A. Lauryl alcohol with 40% EO	8
B. Blend: 83% surfactant A + 17% surfactant G	9
C. Blend: 67% surfactant A + 33% surfactant G	10[1]
D. Blend: 50% surfactant A + 50% surfactant G	11
E. Blend: 33% surfactant A + 67% surfactant G	12
F. Blend: 17% surfactant A + 83% surfactant G	13
G. Blend: Lauryl alcohol with 70% EO	14

[1] Most complete and stable emulsion.

HLB numbers for emulsifying various materials are shown in Table 3.1 [9]. These are usually obtained by simply observing each material's emulsification behavior in association with a series of emulsifying agents having known HLB values. For example, if an emulsion of mineral oil in water is desired, one could formulate mineral oil/water mixtures containing the series of surfactants shown in Table 3.2. Through observation, the HLB of the mineral oil could then be determined based on the surfactant (C) that forms the best (most complete and stable) emulsion. When the HLB number of the substance(s) to be emulsified is known, other surfactants and surfactant mixtures having the same average HLB should be tried to determine the combination that works best.

The following guidelines are useful in selecting surfactants for emulsification.

- Water-soluble surfactants (HLB > 10) are generally best for o/w emulsions, while oil-soluble surfactants (HLB < 10) are best for w/o emulsions.
- Mixtures of surfactants usually form more stable emulsions than a single surfactant, especially if the mixture consists of water-soluble and oil-soluble surfactants.
- Surfactants of different types should be tested, since surfactant types vary in the degree and mechanism by which they stabilize various emulsions.

Other more complicated methods are available to aid in formulating emulsions. These include solubility parameters [10] and phase inversion temperature [11,12].

3.2.4   Stabilize Foam

Foam is generally defined as a coarse dispersion of gas in a relatively small volume of liquid. Some surfactants stabilize foam; some do not. The ability to stabilize foam depends on the ability to increase "elasticity" at the gas–liquid interface. For most systems, elasticity is increased by increasing the level of surfactant present at the air–water interface. The presence of an ionic charge (i.e., anionic and cationic surfactants) and/or oxygen-containing functional groups that can promote hydrogen bonding (e.g., hydroxyl groups) helps to accomplish this. This is why anionic surfactants generally foam better than nonionic surfactants, and why some nonionics (e.g., alkyl polyglycosides) foam better than other nonionics (e.g., alcohol ethoxylates).

Foam is important from an aesthetic point of view for dishwashing liquids, hand soaps, shampoos, and some hard-surface cleaners. Years ago, this was also true of laundry detergents, but problems with rinsing and changing consumer attitudes have led the market to lower foaming ("controlled foam") products.

Foam is also important in industrial processes. In ore flotation, for example, foam is used to separate desired from undesired materials.

3.2.5   Adsorb on Surfaces/Amend Surface Properties

Cationic surfactants, such as quaternary amine salts, irreversibly adsorb onto cloth. This results in neutralization of surface charge (helps to eliminate static charge) and gives cloth a softer feel because the hydrophobic portion of the molecule orients itself away from the cloth. This is why cationic surfactants are used as fabric softeners.

Anionic surfactants do not adsorb onto fabric as strongly as cationic surfactants because cloth normally carries a net negative charge. Nonionic surfactants have not been used as softening agents, although alcohol/alcohol ethoxylate blends have been shown to be effective as antistatic agents [13].

3.2.6   Bactericide

Cationic surfactants are also effective bactericides because they adsorb onto bacteria (cell wall is negatively charged) and irreversibly interfere with processes important to growth and reproduction. Anionic surfactants can inhibit bacterial growth, but because this inhibition is reversible, they are not considered to be bactericides. Nonionic surfactants do not generally act as bactericides.

If you are looking for a bactericide, I recommend contacting suppliers of nitrogen-based surfactants for a discussion of specific needs.

3.2.7   Terminology

Some terms often used in our industry are misused, and it is important to keep this in mind when reading surfactant literature. A "soap" refers to the sodium or potassium salt of a blend of long-chain fatty acids. Soaps are "surfactants," although the latter word (term) is often reserved for surface active agents other than soaps. Similarly, although the word "detergent" should refer to any cleaning agent or formulation, it is often used to designate nonsoap surfactants.

3.2.8   Surfactant Classification by Charge

Surfactants have historically been grouped according to the charge they have when dissociated in water at neutral pH. These categories are:

- *Anionic surfactants*: surfactants that have a negative charge when dissociated in water; normally sold as sodium, potassium, or ammonium salts (e.g., linear alkylbenzenesulfonate, alcohol sulfate, alcohol ether sulfate, soap, secondary alkanesulfonate, α-olefinsulfonate, methyl ester sulfonate).
- *Nonionic surfactants*: surfactants that do not ionize in solution (e.g., alcohol ethoxylates, alkylphenol ethoxylates, ethylene oxide/propylene oxide block copolymers, alkyl polyglycosides).
- *Cationic surfactants*: surfactants that have a positive charge when dissociated in water; normally consist of alkylammonium salts with chloride or sulfate as the counterion (e.g., dodecyltrimethylammonium, ditallowdimethylammonium methosulfate).
- *Amphoteric surfactants*: surfactants that can carry both a positive and negative charge when dissociated in water; also called zwitterionic surfactants (e.g., dodecylbetaine).

3.2.9   Surfactant Classification by Feedstock

Another classification that has recently arisen defines whether the surfactant is "natural" or "synthetic." In reality, these terms do not have any significant meaning since all surfactants, whether based on natural oils, petroleum, or tallow, require extensive chemical processing to manufacture the finished surfactant. Although some surfactant manufacturers tout their surfactants as superior because they are "natural", this is for commercial reasons, not because scientific data shows they are better for the environment. A better distinction that has taken root is "oleochemical" versus "petrochemical." These terms accurately describe the feedstock used to chemically manufacture surfactants, without implying that surfactants derived from "natural" feedstocks are simply plucked from the ground.

3.2.10   Preview of Sections 3.3–3.5

The remainder of this chapter describes various classes of commercial surfactants. It includes surfactants that are currently widely used or have the potential to be so used in the future. There are three main categories, covered in Sections 3.3, 3.4, and 3.5, respectively:

- *Anionic surfactants* (linear alkylbenzenesulfonates, alcohol sulfates, alcohol ether sulfates, soaps, secondary alkanesulfonates, α-olefinsulfonates, and methyl ester sulfonates)
- *Nonionic surfactants* (alcohol ethoxylates, alkylphenol ethoxylates, ethylene oxide/propylene oxide block copolymers, and alkyl polyglycosides)

- *Nitrogen-based surfactants* (cationic, amphoteric, and nonionic surfactants made from amides or amines)

Each section describes what makes up the hydrophilic and hydrophobic portions of the surfactant, how surfactant structure can vary, and how structural variations affect performance. Performance properties and typical applications are also discussed.

3.3   Anionic Surfactants

3.3.1   LAS (Linear Alkylbenzenesulfonate, or Linear Alkylate Sulfonate)

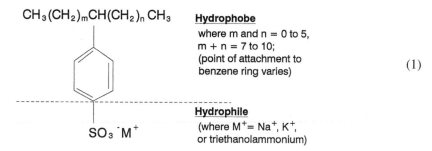

$$CH_3(CH_2)_mCH(CH_2)_nCH_3$$

Hydrophobe
where m and n = 0 to 5,
m + n = 7 to 10;
(point of attachment to
benzene ring varies)

(1)

$$SO_3^- M^+$$

Hydrophile
(where M^+ = Na^+, K^+,
or triethanolammonium)

LAS is the most commonly used surfactant in the world. It has been used for more than 25 years, with almost 5 billion pounds is consumed worldwide every year. LAS is clearly the most studied surfactant because of its large consumption rate and by virtue of the fact that it is easy to detect analytically. This low-priced, high-foaming, all-purpose surfactant formulates well into both powders and liquids. It is used in a variety of household and industrial applications.

3.3.1.1   Description of Hydrophobe

The hydrophobe of LAS is linear alkylbenzene (LAB), sometimes called "detergent alkylate." Manufacture of the LAB is accomplished through Friedel–Crafts alklylation using either chloroparaffins or olefins as the feedstock (Fig. 3.7). Two alkylation catalysts are used commercially: aluminum chloride ($AlCl_3$) and hydrofluoric acid (HF). Solid catalysts are also being developed. Commercial LAS can vary in average molecular weight, phenyl isomer distribution, and dialkyltetralin level. Molecular weight depends on the paraffin or olefin stream used to make the LAB. The isomer distribution is a function of the alkylation catalyst that is used, and the tetralin level is a function of whether chloroparaffins are used for alklylation. The effect of these variations on "hydrophobicity" is discussed below.

3.3.1.1.1   Average Molecular Weight

Commercial LAB is usually made from alkylation of various blends of C_{10}, C_{11}, C_{12}, C_{13}, and C_{14} linear chloroparaffins or olefins. Different average molecular weight LABs are available, normally averaging in the molecular weight range of 232–260 (C_{11}–C_{13}). This specific mole-

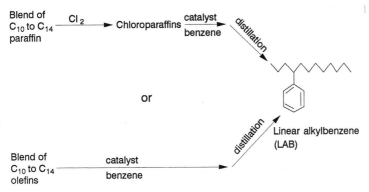

Figure 3.7   Flow diagram showing commercial routes for producing linear alkylbenzene.

cular weight range is the only range that provides both acceptable solubility and surface activity. Increasing molecular weight above 260 produces an LAS with better surface activity but very low water solubility. Decreasing molecular weight below 232 produces a very soluble LAS with poor surface activity. Molecular weight is very important when choosing an LAS, and the proper molecular weight in turn is determined by the conditions under which the LAS is to be used. A lower molecular weight LAS should be used when ionic strength and/or water hardness (level of calcium and magnesium ions) is significant, because both parameters reduce LAS solubility. If ionic strength and water hardness levels are low, a higher molecular weight is better.

3.3.1.1.2   Phenyl Isomer Distribution

During alkylation, the benzene molecule generally attaches itself to any carbon along the alkyl chain (except the terminal methyl groups). The phenyl isomer distribution refers to the concentration of LAB molecules where the benzene group is attached to the second carbon of the alkyl chain, the third carbon of the alkyl chain, the fourth carbon, and so forth. The phenyl isomer distribution depends on the catalyst used during alkylation (Table 3.3). Aluminum chloride alkylation gives what is normally called a "high 2-phenyl" distribution, while HF gives a "low 2-phenyl" distribution. The "2-phenyl" designation is used simply to distinguish between the two major types of alkylate.

The effect of phenyl isomer distribution on water solubility is not straightforward because the solubility of pure isomers depends on temperature, and because isomer solubility is influenced by the presence of other isomers [14]. In general, isomer solubility decreases as phenyl attachment approaches either the end or the center of the alkyl chain. With respect to

Table 3.3   Typical Phenyl Isomer Distributions for HF-Catalyzed and AlCl$_3$-Catalyzed LAB (Commercial C$_{12}$-Average LAB)

Alkylation catalysts	2-Phenyl	3-Phenyl	4-Phenyl	5-Phenyl	6-Phenyl
HF	19	20	20	21	20
AlCl$_3$	29	22	18	16	16

commercial materials, if carbon chain distributions are equal and tetralin level is below 1%, high 2-phenyl LAS is slightly more soluble than low 2-phenyl LAS.

The bottom line is that the phenyl isomer distribution is not particularly important.

3.3.1.1.3   Dialkyltetralin Level

Commercial LAS contains up to 8% dialkyltetralin.

$$(CH_2)_m CH_3$$

where m and n = 0 to 7
where m+n = 4 to 7

(2)

$$(CH_3)_n CH_3$$

The level of dialkyltetralin is determined by the alkylation process. Use of chloroparaffins normally produces alkylate containing 6–8% dialkyltetralin. Chlorination of paraffins produces a significant amount of dichloroparaffins, which can form dialkyltetralin during the alkylation process. Dialkyltetralins readily sulfonate and, when sulfonated, are good surfactants. Dialkyltetralinsulfonates are also extremely good solubilizers, which is why LAB containing 6–8% tetralin is desirable for use in liquids.

Manufacture of olefins produces diolefins (dienes), which also form dialkyltetralin, but diene formation is normally limited to less than 2%. Consequently, LAB made from olefins contains about 1% tetralin.

The environmental acceptability of dialkyltetralins has been questioned based on conjecture that dialkyltetralinsulfonates, which are slow to achieve biodegradation, may be accumulating in the environment. Recent studies have demonstrated, however, that dialkyltetralinsulfonates biodegrade (mineralize) and are not accumulating in the environment [15]. Consequently, dialkyltetralins are not a major issue to the detergent industry.

3.3.1.2   Description of Hydrophile

The hydrophile of LAS is the sulfonate group. The sulfonate group (where sulfur is bonded directly to carbon) should not be confused with the sulfate group (where sulfur is bonded to oxygen which is bonded to carbon). Sulfonation is accomplished using an air–sulfur trioxide (SO_3) mixture or oleum (fuming sulfuric acid). Sulfur trioxide is most often employed because it is more economical and yields a product of better quality. Sulfonation is often done by the end user, although several companies (Stepan Chemical Company, Pilot Chemical Company, Witco Corporation, and others) specialize in sulfonation. With the exception of Vista, most LAB suppliers do not sulfonate their own LAB.

Sulfonation first produces sulfonic acid (Fig. 3.8), which is normally composed of about 96–97% sulfonic acid, 1–2% free oil (sulfone and unsulfonated alkylate), and 1–2% sulfuric acid. The sulfonic acid is stable and can be shipped as is, to be neutralized later by the end user, or it can be neutralized and shipped in the sulfonate form. Sulfonic acid is normally neutralized with a sodium hydroxide solution to produce a 50% active slurry (with approximately 1–2% unsulfonated material, 1% sodium sulfate, and 47–48% water). A 60% active slurry is also available which has 2–3% hydrotope (usually sodium xylene sulfonate) to lower viscosity. Although sodium salts are most common, other bases (potassium hydroxide, triethanolamine, etc.) can be used if sodium cannot be tolerated in the finished formulation. One advantage for shipping sulfonic acid versus the neutralized salt is that less water is shipped, which effectively saves on freight costs.

3.3.1.3   Properties and Characteristics

Physical Properties   LAS is easy to handle and dilute (does not go through a gel phase during dissolution). Reducing molecular weight increases water solubility.

Formulation in Powders   LAS has good powder properties (makes a stable flowable powder), and it can be spray-dried, agglomerated, and dry-neutralized.

Fabric Detergency   LAS is a good overall detergent. Optimum molecular weight depends on use concentration and water hardness level. Under conditions of controllable water hardness [< 25 ppm free water hardness (as $CaCo_3$)], a higher molecular weight LAS is best. When the level of water hardness is significant, a lower molecular weight LAS is best. Increasing ionic strength and/or reducing temperature magnifies this effect.

Wetting   LAS is considered to be a good wetting agent. Optimum molecular weight, like fabric detergency, depends on use conditions. In general, however, a higher molecular weigh LAS is better because of its superior surface activity.

Foaming   LAS stabilizes foam (is considered to be a high foamer). A lower molecular weight is considered best for foaming. LAS is often used in conjunction with alcohol ether sulfate because a blend gives better foaming than either by itself. Fatty amides are often added to dishwashing liquids based on LAS/alcohol ether sulfate to help stabilize the foam.

Compatibility with Additives   LAS is not compatible with aqueous solutions of enzymes without the presence of stabilizers. LAS, because of its negative charge when dissociated, is not compatible with cationic surfactants in aqueous solutions. LAS is also chemically degraded by hypochlorite bleach.

Mildness   LAS is not considered to be harsh, but at high concentrations it can irritate sensitive skin. Consequently, LAS is used in hand dishwashing detergents but not in shampoos.

3.3.1.4   Applications

LAS is most often used in laundry detergents, usually in combination with other surfactants such as alcohol sulfate, alcohol ether sulfate, alcohol ethoxylate, or soap. Mixed active products provide greater versatility with respect to washing conditions. The addition of alcohol ethoxylate and/or fatty acid soap also helps limit foaming. This is important in front-load washers and industrial cleaning equipment.

LAS is also commonly used in hand dishwashing detergents, usually in combination with alcohol ether sulfate (AES). Both LAS and AES are excellent foamers, but blends (particularly in the 4:1 LAS/AES range), provide more stable, soil-tolerant foam. Consequently, all hand dishwashing detergents that contain LAS also contain alcohol ether sulfate.

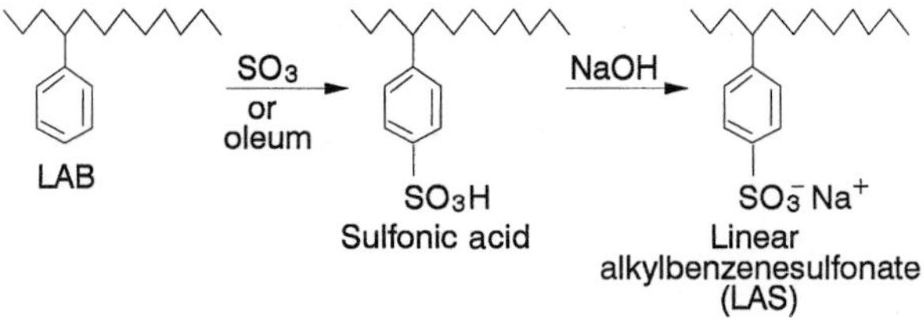

Figure 3.8   Sulfonation of linear alkylbenzene.

LAS is also used in hard-surface cleaners, particularly when foaming, particulate detergency, and powder properties are important.

LAS is not normally used in shampoos or liquid hand soaps because it may be somewhat harsh at the surfactant concentrations typical of products of these types.

3.3.1.5   Availability

LAS is readily available as 50% active and 60% active slurries, and as 90+% active drum-dried flake. It is also often sold in the sulfonic acid form because it is fairly easy to neutralize. In some powder applications, the sulfonic acid can be neutralized by another powder component (e.g., sodium carbonate) to help minimize water removal: this is called "dry neutralization."

LAS is available in different molecular weights. Low molecular weight LAS (330–340) is usually used for liquids, while higher molecular weight LAS (350–360) is usually used in well-built laundry powders.

3.3.2   Alcohol Sulfates (or Alkyl Sulfates)

$$CH_3 (CH_2)_n - \mid O\text{-}SO_3^- \, M^+ \qquad\qquad (3)$$

Hydrophobe | **Hydrophile**

(where n = 11 to 17) | (where M^+ = Na^+, NH_4^+, etc.)

Alcohol sulfates are one of the three major anionic workhorse surfactant types (along with LAS and alcohol ether sulfates). Alcohol sulfates have been used extensively for more than 25 years in a wide variety of applications. They are high-foaming, efficient, moderate-priced surfactants with excellent detergency properties. Solubility and performance, however, are affected by water hardness ions, ionic strength, and temperature, and these surfactants are somewhat harsh on skin. More than 750 million pounds of alcohol sulfates is consumed annually worldwide.

3.3.2.1   Description of Hydrophobe

The hydrophobe of an alcohol sulfate is the alkyl chain of the alcohol. Most alcohols are produced by two petrochemical and two oleochemical routes. The two predominant petro-

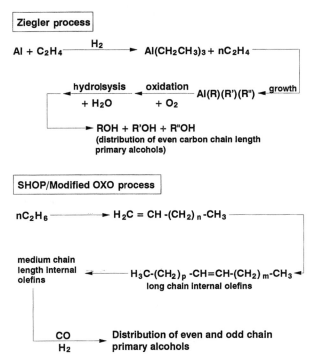

Figure 3.9 Major petrochemical routes for making alcohols.

chemical routes are the Ziegler process and the SHOP/modified OXO process (Fig. 3.9). The Ziegler process involves the use of aluminum metal to grow alkyl chains from ethylene prior to oxidation and hydrolysis to produce a distribution of primary alcohols that are even in carbon chain length. This chemistry is employed by Vista, Ethyl, Condea (Germany), and Ufa (Russia). Shell Chemical Company utilizes its SHOP (Shell Higher Olefin Process) to produce first even-carbon-chain α-olefins and then internal olefins of even and odd carbon chain length. Shell next uses a modified OXO process to covert the internal olefins to alcohols. (The OXO process involves conversion of α-olefins to alcohols).

Oleochemical alcohol production can utilize a variety of oils, but generally coconut and palm kernel oils are used because they are highly saturated and produce mostly alcohols in the lauryl (C_{12} and C_{14}) range (most desirable). There are basically two routes used for converting oils (triglycerides) into alcohols (Fig. 3.10). The most prevalent route involves transesterification of the triglyceride to methyl esters, which are then hydrogenated to alcohols. In the other route, the oil is converted to fatty acids followed by hydrogenation of the acids to alcohols. Both routes have process advantages and disadvantages. Both processes produce high quality primary alcohols of even carbon chain length.

Approximately half the alcohol used to make alcohol sulfate comes from oleochemical feedstocks. Although suppliers often debate the advantages of using petrochemical and oleochemical feedstocks, in reality there is not a great deal of difference in quality and composition between the two with the exception of (a) the degree of methyl branching along the alkyl chain and (b) with respect to whether the alcohol contains odd-numbered carbon chain lengths. As shown in Table 3.4, even these differences are related less to the source of feedstock than to the manufacturing process used to make the alcohol.

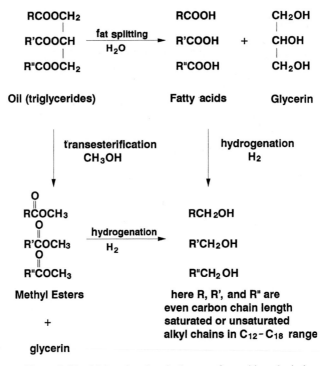

Figure 3.10 Major oleochemical routes for making alcohols.

Table 3.4 Composition of Oleochemical and Petrochemical-Based Alcohols

Feedstock/process	Alcohol type	Even vs. odd chains	Methyl branching (%)
Oleochemical	Primary	Even	0
Petrochemical/Ziegler	Primary	Even	~ 2
Petrochemical/modified OXO	Primary	Even and odd	~ 20

Suppliers have also debated the benefits of odd versus even carbon chain lengths. It is not very important, however, whether the alcohol contains even or even and odd carbon chains. What is important is the average molecular weight and, to a smaller extent, the degree of methyl branching. The impact of methyl branching is relatively minor. A higher degree of branching results in a lower solution viscosity in the end product and can give slightly higher sulfation color.

In terms of average molecular weight, or alkyl chain length, most alcohol sulfates are made from various blends of alcohols in the C_{10}–C_{18} range. Increasing carbon chain length increases hydrophobicity and decreases water solubility. Tallow-range (C_{16} and C_{18}) alcohol sulfate is an excellent surfactant, but it is not very soluble. Lauryl-range (C_{12} and C_{14}) alcohol sulfate is less surface active but it is more soluble.

3.3.2.2   Description of Hydrophile

The hydrophile of alcohol sulfate is the sulfate group. The sulfate group (where oxygen bridges the sulfur and the carbon of the alkyl chain) should not be confused with the sulfonate group (where sulfur is bonded directly to carbon). Sulfation (Fig. 3.11) is straightforward and can be accomplished using sulfur trioxide, oleum (SO$_3$ dissolved in sulfuric acid), sulfamic acid, and chlorosulfonic acid. Sulfur trioxide is generally preferred by the end user because it typically produces the best quality sulfate in terms of completeness of sulfation and color. Oleum and chlorosulfonic acid are also used: oleum when color and salt level are less important; chlorosulfonic acid when sulfation cost and salt level are less important. Neutralization is usually accomplished with sodium or ammonium hydroxide, although other agents (triethanolamine, potassium hydroxide, etc.) can be used.

Although sulfation is straightforward, minor differences in the completeness of sulfation and the amount of salt produced can affect how well the solution accepts being formulated. Appreciable amounts of unsulfated material (free oil) and/or inorganic salts result in higher solution viscosity [16]. This is why consistency in supply is important to many end users.

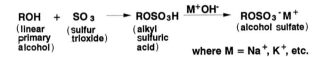

$$ROH \; + \; SO_3 \longrightarrow ROSO_3H \xrightarrow{M^+OH^-} ROSO_3{}^-M^+$$

(linear primary alcohol)   (sulfur trioxide)   (alkyl sulfuric acid)   (alcohol sulfate)

where M = Na$^+$, K$^+$, etc.

Figure 3.11   Sulfation of linear alcohols.

3.3.2.3   Properties and Characteristics

Physical Properties   Lauryl-range alcohol sulfates gel at concentrations above about 28%, which is why they are normally sold in 25–28% solutions. Tallow-range alcohol sulfates are less soluble and gel at lower concentrations.

Formulation in Powders   Alcohol sulfates are often incorporated into powders. In comparison to sulfonates, however, sulfates are more susceptible to thermal and hydrolytic degradation problems.

Fabric Detergency   Alcohol sulfates are excellent detergents (especially on particulate-type soils), although they are sensitive to temperature, ionic strength, and water hardness. To maintain optimum detergency, average molecular weight of the alcohol sulfate needs to decrease with decreasing wash temperature and/or increasing water hardness and ionic strength.

Wetting   Alcohol sulfates are good general-purpose wetting agents and are often used when foaming is important.

Foaming   Alcohol sulfates are known for the quantity and quality of the foam they produce. Alcohol sulfate foam is dense and aesthetically pleasing to look at and feel. Lauryl-range alcohol sulfates are usually used in foaming applications instead of tallow-range alcohol sulfates because solubility is normally important.

Compatibility with Additives   In solution, alcohol sulfates are not compatible with cationic surfactants or enzymes. In contrast to most anionics, alcohol sulfates do show some tolerance for hypochlorite bleach.

Mildness   Alcohol sulfates are known to be somewhat harsh on skin. However, with the inclusion of additives, they are still formulated into personal care and hand dishwashing products.

3.3.2.4   Applications

The majority of alcohol sulfates end up in detergent powders in combination with other surfactants. Several decades ago, tallow-range alcohol sulfates were used in detergent powders because the consumer liked to see lots of foam and used hot water; moreover, detergent formulations contained high levels of phosphate to control the effects of water hardness ions. As phosphate usage and wash temperatures declines, tallow-range alcohol sulfate becomes less acceptable. Many modern detergents still use alcohol sulfate, but normally in combination with other surfactants (LAS, etc.), to improve overall tolerance to wash conditions and to control (reduce) foam.

   Although alcohol sulfates are still used for personal care and hand dishwashing formulations, they have largely been replaced by alcohol ether sulfates, which have similar foaming properties but are less harsh on the skin. Alcohol sulfates are also used in some hard-surface cleaners, particularly when foaming is important. Very pure lauryl-range alcohol sulfate (usually sold as drum-dried powder) is also used in toothpastes.

3.3.2.5   Availability

Alcohol sulfates are generally available as 28–30% solutions, and some are available as 90+% powders. Alcohol sulfate is not available in the acid form because it is unstable.

3.3.3   Alcohol Ether Sulfates (AES) or Alkyl Ether Sulfates or Alcohol Ethoxysulfates

$$CH_3 (CH_2)_x - \quad O(CH_2CH_2O)_nSO_3^- M^+ \qquad (4)$$

Hydrophobe   **Hydrophile**
(where $x = 11$ to 17)   (where $n=0$ or higher depending
                                on degree of ethoxylation;
                                $M^+ = Na^+$ or NH_4^+)

Alcohol ether sulfates represent the third workhorse anionic surfactant. They have also been used for more than 25 years in a variety of applications including shampoos, dishwashing liquids, laundry detergents, and wallboard (gypsum board) manufacture. They are high-foaming, moderate-priced surfactants that are milder to skin than alcohol sulfates and LAS. As detergents, alcohol ether sulfates do not generally perform as well as alcohol sulfates, but they are less affected by water hardness, ions, and temperature. Alcohol ether sulfates are also unique in comparison to other anionic surfactants in that both the hydrophile and hydrophobe portions of the molecule can be modified to give a range of physical and surface chemical properties.

   Approximately 1 billion pounds of alcohol ether sulfates is consumed annually worldwide.

3.3.3.1   Description of Hydrophobe

The hydrophobe of an alcohol ether sulfate is the alkyl chain of a linear alcohol, just like an alcohol sulfate. Alcohols used to make AES are often identical to those used to produce alcohol sulfates, originating from either oleochemical or petrochemical feedstocks (see Section 3.3.2). As with alcohol sulfates, an increase in alkyl chain length increases hydrophobicity (makes the surfactants less water soluble and more surface active). Lauryl-range alcohols are generally best for solubility, foaming, and salt thickening (see below). Higher molecular weight alcohols are used when surface properties (detergency, etc.) are most important.

3.3.3.2   Description of Hydrophile

The hydrophile of alcohol ether sulfate consists of the ethylene oxide chain and the sulfate group. Ethoxylation and sulfation are performed independently (Fig. 3.12), usually by different companies. Ethoxylation of alcohols is discussed elsewhere (see Section 3.4.1). Sulfation of alcohol ethoxylates is similar to alcohol sulfates (see Section 3.3.2).

Length of the ethylene oxide (EO) chain can vary significantly, yielding surfactants containing anywhere from 0 to more than 10 units of EO. For most applications, however, the average number of EO units varies between 1 and 3 moles (units) of ethylene oxide. At these levels of ethoxylation, the amount of nonethoxylated alcohol is significant and can impact performance, as discussed below.

Ethoxylation of an alcohol yields a distribution of alcohol ethoxylates that varies in the number of ethylene oxide units making up the EO chain. Some of the alcohol has one unit of EO attached, some has two units, some has three units, and so on. A portion of the alcohol also remains free (unethoxylated). Table 3.5 gives the distribution of three alcohol ether sulfates containing an average of 1.1., 1.8, and 2.9 moles of ethylene oxide on dodecyl (C_{12}) alcohol. In all three cases, the largest single component is the unethoxylated alcohol. As more EO is added, the amount of unethoxylated alcohol decreases. When these ethoxylates are sulfated, the unethoxylated alcohol is converted to alcohol sulfate. Alcohol ether sulfates having less than 4 moles of EO, therefore, contain a significant amount of alcohol sulfate.

The effect of EO chain length on AES performance is less straightforward than with alcohol ethoxylates because it impacts both the hydrophilicity and the hydrophobocity of the molecule. Although increasing EO chain length increases water solubility (associated with an increase in hydrophilic character), it also increases aggregation number and critical micelle concentration, which are associated with an increase in hydrophobicity [17,18].

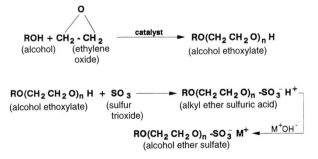

Figure 3.12   Preparation of an alcohol ether sulfate.

Table 3.5 Composition (wt %) of 1.1-, 1.8-, and 2.9-Mole Alcohol
Ether Sulfates Made From Dodecyl (C_{12}) Alcohol

	1.1-Mole AES	1.8-Mole AES	2.9-Mole AES
Free alcohol	42	29.2	15.5
1-Mole	19	15.7	11.1
2-Mole	12	12.6	11.5
3-Mole	9	11.2	11.5
4-Mole	6	8.7	10.3
5-Mole	4	6.3	8.5
6-Mole	3	4.9	7.4
7-Mole	2	3.8	6.3
8-Mole	1	2.7	5.1
9-Mole	<1	1.9	4.0
10-Mole		1.3	2.9
11-Mole		<1	2.1
12-Mole			1.4
13-Mole			<1
Average wt. % EO:	21.0	29.8	40.9

3.3.3.3   Properties and Characteristics

Physical Properties   Lauryl-range alcohol ether sulfates are very water soluble and formulate well into liquids. Dilution of AES, however, can be difficult because AES goes through a gel phase during dilution. This is why it is better to add AES to water than vice versa. Tallow-range alcohol ether sulfates are less soluble and more difficult to salt-thicken (see below); they form gels at lower concentrations (ca. 22–24%) in comparison to lauryl-range AES (ca. 27–29%).

Formulation in Powders   Alcohol ether sulfates are somewhat hygroscopic and are used in powders only when other less hygroscopic surfactants (LAS, alcohol sulfates) are also incorporated.

Salt Thickening   A key property of alcohol ether sulfates is their ability to salt-thicken. This allows the formulator to produce highly viscous formulations, which the consumer equates with quality and performance. At moderate concentrations, AES is present mostly in the form of spherical micelles. Micelles are spherical because a sphere maximizes the distance between the negatively charged head groups, which repel each other. Micelle size is generally fixed because once a micelle has grown large enough, it has sufficient negative charge to repel other molecules. If ionic strength is increased (salt is added), the negatively charged head groups can be surrounded by cations. This effectively shields the negative charge and allows the micelle to grow in size and form cylindrical micelles. Formation of these cylindrical micelles increases solution viscosity dramatically.

The effect of AES composition on salt thickening is discussed elsewhere [17]. In summary, both EO chain length and alkyl chain length are important. A 1-mole AES salt-thickens better than higher mole alcohol ether sulfates. A lauryl-range AES salt-thickens better than a tallow-range AES.

Fabric Detergency   Like LAS and alcohol sulfates, alcohol ether sulfates are good detergents, especially on particulate soils. Although AES does not produce as stable and crisp a powder in comparison to LAS and alcohol sulfates, it is less sensitive to temperature and water hardness effects. Overall, an average chain length of between C_{14} and C_{16} with 2–4 moles of ethylene oxide is considered to give maximum performance.

Alcohol ether sulfates have also been to decrease the sensitivity of LAS to water hardness by helping solubilize LAS–water hardness complexes [19].

Wetting   Alcohol ether sulfates are good general-purpose wetting agents and are used when foaming and solubility are important.

Foaming   Alcohol ether sulfates are known for their ability to stabilize foam. Although AES foam is generally not considered to be quite as rich and dense as alcohol sulfate foam, it is preferred in skin contact applications (because AES is milder to skin), in industrial applications (because it is more tolerant to soil, water hardness, and temperature), and in applications requiring moderate to low concentrations of surfactant (because AES foams better than alcohol sulfate when concentrations are low). Maximum foaming is normally observed with lauryl-range alcohols, although octanol and decanol also produce high-foaming surfactants. Optimum EO content depends on use concentration. At high concentrations (> 10,000 ppm), alcohol sulfate and 1-mole AES are about equal in terms of their ability to generate foam and with respect to how well the foam of each can tolerate soil [17]. Both produce foams substantially more stable than higher EO-containing AESs. At lower concentrations, optimum foam stability is observed with 1–3 moles of EO.

AES is also known for its foam stability synergism with LAS. This is why dishwashing liquids containing LAS always contain AES.

Compatibility with Additives   Alcohol ether sulfates are more compatible with detergent enzymes than alcohol sulfate and LAS. AES is less compatible than alcohol sulfate to bleach, presumably because of the ether linkage between the alkyl chain and the ethylene oxide chain. Like LAS and alcohol sulfates, AES binds with cationic surfactants. The resulting complex, however, is more stable.

Mildness   Alcohol ether sulfates are milder than alcohol sulfates and LAS [20]. Since the degree of ethoxylation determines the amount of alcohol sulfate present in the AES, it follows that a longer EO chain results in a greater degree of mildness.

The Effect of "Peaking" the EO Distribution on Performance   Some ethoxylators also manufacture "peaked" or "narrow range" ethoxylates (see Section 3.4.1). Peaking affects AES performance in two ways:

- Peaking decreases the amount of unethoxylated alcohol present in the parent ethoxylate; AES made from a peaked ethoxylate therefore contains less alcohol sulfate and thus produces a milder surfactant.
- Peaking increases the ability of the surfactant to be salt-thickened [21,22].

3.3.3.4   Applications

Alcohol ether sulfates are used in laundry detergent powders (higher molecular weight 2–3 moles AES), laundry liquids (lauryl-range, 3–6 moles AES), dishwashing liquids (lauryl-range, 1–3 moles), shampoos (lauryl-range 1–3 mole AES), and wallboard manufacturing (C_8–C_{10} range, 2–3 moles AES). With the exception of shampoos and wallboard manufacture, AES is often used in conjunction with other surfactants because of its relatively high cost.

3.3.3.5   Availability

Alcohol ether sulfates are generally available as 28–30% solutions, 60% solutions that contain hydrotrope (ethanol, sodium xylene sulfonate), and 70% pastes (in Europe). AES is not available in the acid form because it is unstable.

3.3.4   Soap (Sodium or Potassium Salts of Long-Chain Fatty Acids)

$$R\text{-}\overset{\displaystyle O}{\overset{\|}{C}}\text{-}O^- M^+ \tag{5}$$

Hydrophobe
(C_{11} -C_{17} linear alkyl chain)

Hydrophile
(where M^+= Na^+ or K^+)

Soaps have been around a long time, possibly for thousands of years, and for centuries were the sole surfactant used to clean clothes and wash humanity. Today, soaps continue to dominate body cleansing and are still used to some extent in laundry products. Approximately 6 billion pounds of various soaps is consumed annually worldwide.

3.3.4.1   Description of Hydrophobe

The hydrophobe of soap is the alkyl chain of a long-chain fatty acid. Fatty acids are produced from fats (solids at room temperature) or oils (liquids at room temperature). Fats and oils are "triglycerides," which are made up of three fatty acids connected to one molecule of glycerol via ester linkages. Triglycerides are "saponified" or "split" to form three units of fatty acid and one unit of glycerin (Fig. 3.13). Any fat or oil can be used to make soap. The origin of the fat or oil determines the degree of unsaturation and the carbon chain length distribution of the fatty acids. Since unsaturation and carbon chain length affect performance, commercial operations use blends of various feedstocks, most often coconut oil and high quality tallow (animal fat).

3.3.4.2   Description of Hydrophile

The hydrophile of soap is the carboxylate salt (usually sodium or potassium) portion of the neutralized fatty acid. Neutralization can be done after the fatty acids have been refined and purified, or splitting and neutralization can be done in the same batch process.

3.3.4.3   Properties and Characteristics

Physical Properties   Soaps are easily extruded, molded, and milled if water content is kept below about 15%. Good quality (low impurities), high saturation, high average molecular

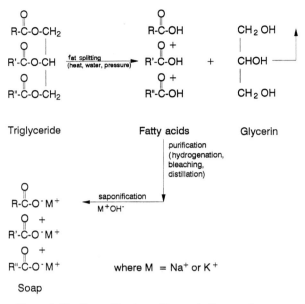

Figure 3.13   Saponification of fats and oils to make soap.

weight sodium salts produce more solid soaps. In contrast, impurities, high unsaturation, low molecular weight potassium salts are more liquid.

Formulation in Powders   Soaps can be flaked after drum-drying. In some markets, drum-dried soap flakes continue to be boxed and sold as is for laundry use.

Fabric Detergency   Soaps are good detergents but are notorious for the water-insoluble precipitate they form with water hardness ions. Formation of soap scum (calcium and magnesium salts of long-chain fatty acids) effectively removes the surfactant from solution and leads to unsightly deposits on clothing. This problem motivated the chemical industry to develop synthetic surfactants. Soap is still used in laundry detergents, but as an adjuvant to help reduce foaming, aid in hardness removal, and help with fabric softening.

Foaming   Although soap is not exceptional in stabilizing foam, it does set the standard for hand lathering. Free fatty acid (up to 10%) increases lathering and improves skin feel. When added to anionic surfactants, soap acts as a soil that disrupts foam stability.

Compatibility with Additives   Soap is not compatible with aqueous solutions of enzymes, with bleaches, or with cationic surfactants.

Mildness   Soap is considered to be about as mild as other anionic surfactants.

3.3.4.4   Applications

Soap is generally used in soap bars and as an adjuvant in laundry detergents.

3.3.4.5   Availability

Different grades of soap are available as solids, usually in the form of flakes.

3.3.5   Secondary Alkanesulfonates (SAS)

$$CH_3\,(CH_2\,)_m\text{-}\overset{\displaystyle H}{\underset{\displaystyle CH_3(CH_2)_n}{C}}\text{-}\;SO_3^-M^+ \tag{6}$$

Hydrophobe
(where m and n=0 to 13
and m+n=8 to 13)

Hydrophile
(where M^+=Na$^+$,etc.)

Secondary alkanesulfonates have been used for more than a decade, predominantly in liquid products, because of their superior water solubility and excellent foaming properties. Their use has been limited because of difficulties in handling and shipping. SAS is used almost exclusively in Europe and is available in both industrial and detergent grades. Approximately 200 million pounds of secondary alkanesulfonates is consumed worldwide annually.

3.3.5.1   Description of Hydrophobe

The hydrophobe of secondary alkanesulfonates is a blend of normal paraffins in the C_{13}–C_{18} homolog range. To produce material of acceptable quality, the paraffin must be low in unsaturation, aromatics, and so on. Paraffins are derived from petrochemical feedstocks.

3.3.5.2   Description of Hydrophile

Sulfonation of normal paraffins is technically difficult. It involves a light-initiated sulfoxidation process (Fig. 3.14). Because yield is poor, extensive separation and paraffin recycling are required. Even at low yields, 5–10% of the active ingredient consists of various disulfonates. Attachment of the sulfonate group occurs at all secondary positions along the paraffin chain (primary sulfonates are not produced). A large by-product stream of spent sulfuric acid is also produced.

3.3.5.3   Properties and Characteristics

Physical Properties   SAS is known for its superior liquid solubility. Formulations that normally require hydrotrope often do not need hydrotrope with SAS. Although increasing alkyl carbon chain length decreases solubility, even the C_{18} homolog is relatively water soluble.

Formulation in Powders   SAS is not generally formulated into powders because it is not available in the acid form and is expensive as a 90+% active surfactant. Intermediate molecular weight SAS (C_{14}–C_{16} SAS) gives good detergency performance which is less affected by water hardness and temperature than are alcohol sulfate and LAS.

Wetting   Increasing molecular weight increases overall wetting properties. SAS is recommended as a wetting agent when foaming and bleach compatibility are important.

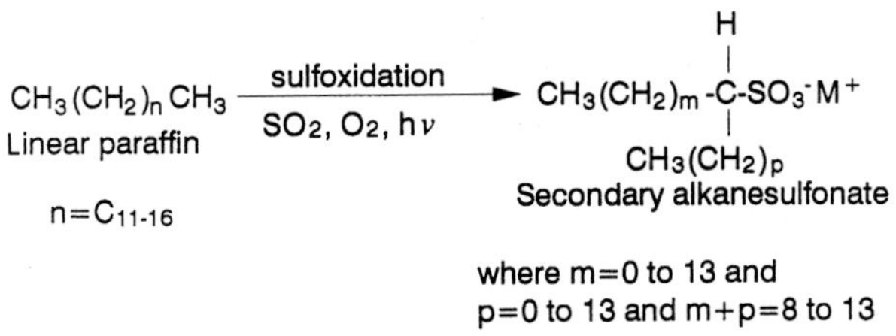

$$\underset{\substack{\text{Linear paraffin} \\ n=C_{11\text{-}16}}}{CH_3(CH_2)_n\,CH_3} \xrightarrow[\text{SO}_2,\,\text{O}_2,\,h\nu]{\text{sulfoxidation}} \underset{\substack{CH_3(CH_2)_p \\ \text{Secondary alkanesulfonate}}}{CH_3(CH_2)_m\overset{\overset{\textstyle H}{|}}{-\!C\!-}SO_3^-\,M^+}$$

where m=0 to 13 and
p=0 to 13 and m+p=8 to 13

Figure 3.14   Preparation of secondary alkanesulfonate.

Foaming   Secondary alkanesulfonates are excellent foamers and compete effectively with LAS/alcohol ether sulfate and alcohol sulfate/alcohol ether sulfate surfactant blends. Secondary alkanesulfonate based formulations generally do not require the addition of foam boosters to help stabilize foam in the presence of soil.

Compatibility with Additives   SAS shows surprising stability with hypochlorite bleach. Like other anionics, however, it is not compatible with cationic surfactants and enzymes.

Mildness   Secondary alkanesulfonates are considered about as mild as other sulfonates.

3.3.5.4   Applications

SAS is used in industrial applications as a water-soluble, general-purpose wetting agent and in dishwashing liquids.

3.3.5.5   Availability

SAS is available as a 30% solution, a 60% slurry, and a 90$^+$% solid.

3.3.6   α-Olefinsulfonate (AOS)

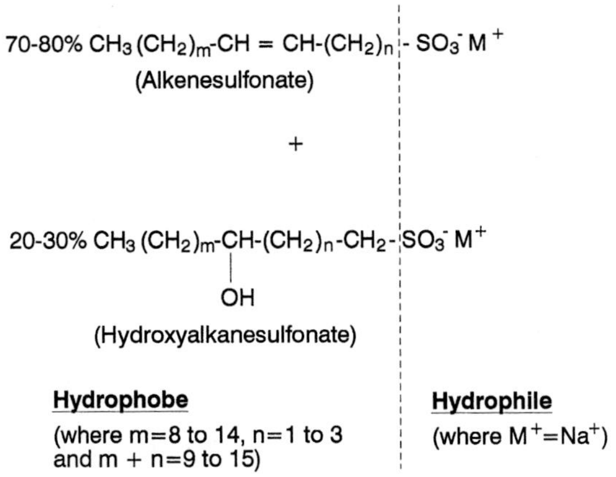

70-80% $CH_3(CH_2)_m\text{-}CH = CH\text{-}(CH_2)_n\text{-}SO_3^-\,M^+$
(Alkenesulfonate)

+

$$\underset{\substack{|\\ OH \\ \text{(Hydroxyalkanesulfonate)}}}{20\text{-}30\%\ CH_3(CH_2)_m\text{-}CH\text{-}(CH_2)_n\text{-}CH_2\text{-}SO_3^-\,M^+}$$ (7)

Hydrophobe   **Hydrophile**
(where m=8 to 14, n=1 to 3   (where M$^+$=Na$^+$)
and m + n=9 to 15)

α-Olefinsulfonates have been around for several decades, and at one time, were thought to be a potential replacement for LAS. Development of AOS as a workhorse surfactant, however, has been hampered by inconsistent quality, concerns about the potential of impurities to cause skin irritation, and the absence of significant advantages of AOS over the more readily available LAS.

AOS is a high-foaming surfactant with good detergency and wetting properties. Approximately 200 million pounds is used worldwide in liquid hand soaps, other personal care products, dishwashing liquids, and laundry detergents.

3.3.6.1   Description of Hydrophobe

The hydrophobe of AOS is an α-olefin. Only α-olefins from petrochemical feedstocks (ethylene) are used because of their high purity. Commercial AOS consists of a blend of C_{14}, C_{16}, and C_{18} olefin, which average in the C_{14}–C_{15} range. This somewhat narrow distribution gives the best combination of low cost, good water solubility, and good surface properties.

3.3.6.2   Description of Hydrophile

Sulfonation of α-olefins is less straightforward than that of linear alkylbenzenes (Fig. 3.15). Sulfonation produces a mixture of hydroxyalkanesulfonates (3- and 4-hydroxyl) and alkenesulfonates (1-alkenesulfonate, 2-alkenesulfonate, etc.), roughly in a 30:70 ratio.

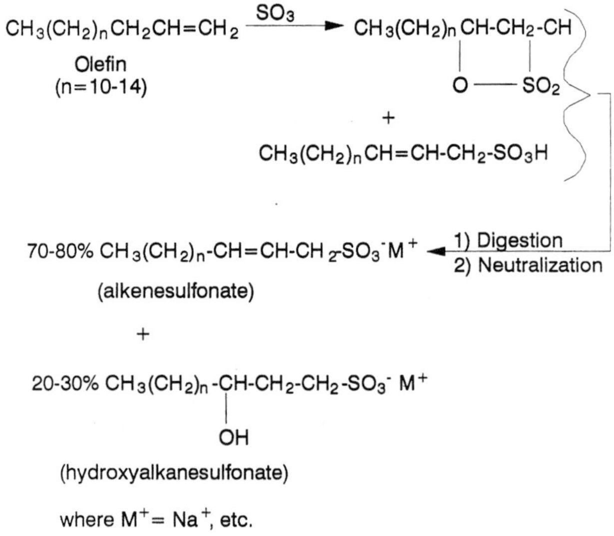

Figure 3.15   Preparation of α-olefinsulfonate.

3.3.6.3   Properties and Characteristics

Physical Properties   AOS is very soluble in water and easy to dilute. It does, however, form a viscous gel when the surfactant concentration exceeds approximately 40%. Increasing the average alkyl carbon chain length decreases solubility.

Formulation in Powders   AOS has powder properties similar to those of LAS.

Fabric Detergency/Wetting/Foaming/Compatibility   AOS is comparable to LAS in terms of detergency, wetting, foaming, and compatibility. It lathers well and was once the predominant active ingredient in liquid hand soaps.

Mildness   Poor quality AOS can contain trace quantities of sultones. When AOS is bleached with hypochlorite, chlorosultones may form, possibly resulting in skin irritation (sensitization) problems. High quality AOS, however, is not considered to be a source of skin irritation.

3.3.6.4   Applications

AOS was first introduced on a country-wide basis in a liquid hand soap. Although it is used now in a variety of personal care products, it is not the principal surfactant in any major laundry detergent or hand dishwashing liquid.

3.3.6.5   Availability

AOS is generally available as a 40% active solution. It is also available in a 90+% (drum-dried) form. The sulfonic acid form is not available because of stability problems.

3.3.7   Methyl Ester Sulfonate (MES)

$$CH_3(CH_2)_n - \overset{\displaystyle O}{\overset{\displaystyle \|}{\underset{\displaystyle \underset{\textstyle SO_3^- Na^+}{|}}{CH\text{-}C\text{-}O}}} X \tag{8}$$

Hydrophobe   **Hydrophile**
(where n=9-15)   (where x= CH_3 for MES and
                                      Na^+ for the disodium salt)

Methyl ester sulfonates have been around for a number of years. Although they are attractive because they utilize an oleochemical feedstock, several problems hindered their development, and they have not captured significant market share. For example, processing problems in initial commercial productions led to material of poor and inconsistent quality. Thermal and hydrolytic stability problems also make it more difficult to formulate MES in laundry powders in comparison to LAS.

3.3.7.1   Description of Hydrophobe

The hydrophobe of MES is the alkyl chain of a methyl ester of a long-chain (C_{12}–C_{18}) fatty acid. Methyl esters have to be hydrogenated to minimize by-product formation. Major sources of MES are palm kernel oil, palm oil, and coconut oil. Methyl esters are mostly used to produce alcohols but could be readily available if demand for MES increased.

Increasing molecular weight increases surface activity but decreases solubility. Tallow-range MES is recommended for laundry detergents while lauryl-range MES is recommended for liquid and foaming applications.

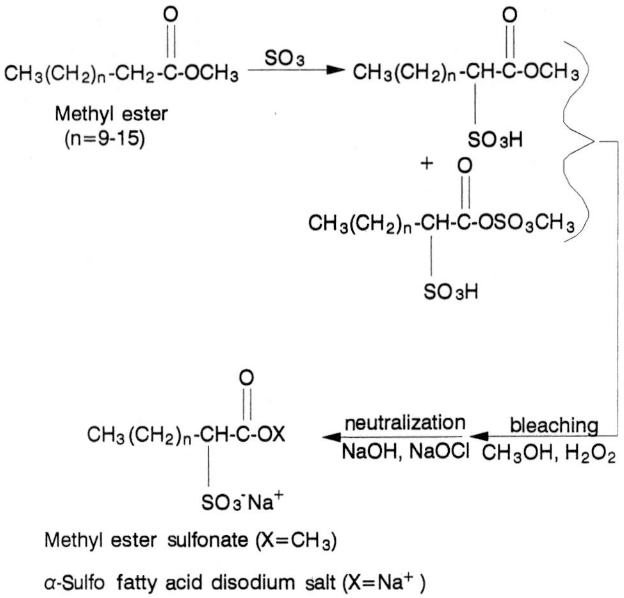

$$CH_3(CH_2)_n\text{-}CH_2\text{-}\underset{\displaystyle \overset{\displaystyle O}{\|}}{C}\text{-}OCH_3 \xrightarrow{SO_3} CH_3(CH_2)_n\text{-}\underset{\displaystyle SO_3H}{CH}\text{-}\underset{\displaystyle \overset{\displaystyle O}{\|}}{C}\text{-}OCH_3$$

Methyl ester
(n=9–15)

$$+$$

$$CH_3(CH_2)_n\text{-}\underset{\displaystyle SO_3H}{CH}\text{-}\underset{\displaystyle \overset{\displaystyle O}{\|}}{C}\text{-}OSO_3CH_3$$

$$CH_3(CH_2)_n\text{-}\underset{\displaystyle SO_3^-\,Na^+}{CH}\text{-}\underset{\displaystyle \overset{\displaystyle O}{\|}}{C}\text{-}OX \xleftarrow[\text{NaOH, NaOCl}]{\text{neutralization}} \xleftarrow[\text{CH}_3\text{OH, H}_2\text{O}_2]{\text{bleaching}}$$

Methyl ester sulfonate (X=CH$_3$)

α-Sulfo fatty acid disodium salt (X=Na$^+$)

Figure 3.16   Preparation of methyl ester sulfonate.

3.3.7.2   Description of Hydrophile

Sulfonation of methyl esters (Fig. 3.16) is difficult and complex. Sulfonation requires aging to reach adequate completion. Bleaching, required to get acceptable color, is usually done prior to neutralization. Hydrolysis of the methyl ester sulfonic acid occurs in all steps and produces an α-sulfo fatty acid disodium salt. Disodium salt content can be minimized by adding excess methanol during the bleaching step [23]. Recent technology allows for production of essentially 100% MES (low levels of the disodium salt) with approximately 1–2% unsulfonated methyl ester.

3.3.7.3   Properties and Characteristics

Physical Properties   MES, like other anionic slurries, is easy to handle and dilute. Solubility decreases significantly with increasing fatty acid chain length.

Formulation in Powders   Development of MES has been hindered by problems with thermal and hydrolytic stability, which have made introduction of MES into spray-dried laundry detergents more difficult. These problems, however, can be overcome through adjustments in processing parameters (temperature, etc.) and by efforts to minimize contact of MES with highly alkaline materials (adjusting pH, utilizing alternate builder systems, etc.).

Fabric Detergency   Tallow-range MES is considered to be a good detergent. It is more tolerant of water hardness than high molecular weight LAS and the higher molecular weight alcohol sulfates.

Wetting/Foaming   MES is considered to be a good foamer. Lauryl-range MES foams better than tallow-range MES.

Compatibility with Additives   MES is reported to be more compatible with enzymes than LAS or alcohol sulfates. It is sensitive, however, to pH (optimum pH is in the 6–9 range). Like other anionics, MES forms insoluble complexes with cationic surfactants and is unstable with hypochlorite bleach.

Mildness   Lauryl-range MES is reported to be mild [24].

3.3.7.4   Applications

In the United States, lauryl-range MES is being marketed for a variety of applications, including heavy-duty liquids and dishwashing liquids [24]. In Japan, MES in the C_{14}–C_{16} range is being used in laundry detergents.

3.3.7.5   Availability

Lauryl-range MES is available as a 40% solution. Tallow-range MES is not commercially available. The sulfonic acid cannot be shipped in the acid form because of poor stability.

3.4   Nonionic Surfactants

3.4.1   Alcohol Ethoxylates, or Alkylpoly(ethylene Oxides), or Alkyl-Poly Oxytheylene Ethers

$$CH_3 (CH_2)_x \text{ - } \mid O(CH_2CH_2O)_nH \tag{9}$$

Hydrophobe   **Hydrophile**
(where x = 5-17)   (where n = 0 or higher
                       depending on degree of
                       ethoxylation)

More than 1.7 billion pounds of alcohol ethoxylates is consumed annually worldwide, making alcohol ethoxylates by far the most widely used nonionic surfactant. Alcohol ethoxylates are relatively inexpensive, low-to-moderate foaming surfactants with excellent wetting and detergency properties, particularly on oily soils. They are used in a wide variety of applications. Like alcohol ether sulfates, both the hydrophobe and the hydrophile can be adjusted to yield a wide variety of molecules.

3.4.1.1   Description of Hydrophobe

Like alcohol sulfates and alcohol ether sulfates, the hydrophobe of an alcohol ethoxylate is normally a linear, primary alcohol. Since ethoxylation involves very reactive chemistry, there is no difference between ethoxylates made with oleochemical and petrochemical feedstocks (see Section 3.3.2). Most major ethoxylators use petrochemically derived alcohols because of availability and cost. Secondary alcohols (derived from petrochemical feedstocks) are also used to produce ethoxylates with enhanced solubility.

Alkyl carbon chain length can vary significantly (from C_6 to C_{18}) but usually consists of alcohol blends in the short-chain range (C_6–C_{10}), the lauryl range (C_{12}–C_{14}), and the tallow range (C_{16}–C_{18}). Increasing chain length increases hydrophobicity. Optimum carbon chain length depends on the application, use concentration, and EO content.

3.4.1.2   Description of Hydrophile

The hydrophile of an alcohol ethoxylate is a chain made up of ethylene oxide units. The average length of the EO chain can vary significantly (< 0–100 moles of EO) depending on what is required. The average amount of EO added to the alcohol (or length of the EO chain) is commonly expressed either as the average weight percent of EO in the molecule or as the number moles (units) of EO making up the EO chain. There are advantages to both systems. Weight percent gives the general ratio of hydrophile to hydrophobe, regardless of alcohol, which is useful in quickly classifying water solubility. Ethoxylates having greater than 50% EO are water soluble; those having less than 50% are oil soluble, and 50% ethoxylates are borderline. Weight percent also allows HLB to be estimated by dividing the weight percent by 5 (e.g., a 60% ethoxylate has an HLB of 12). Moles give a better indication of composition by describing exactly how many units of EO, on average, have been added to the alcohol.

Keep in mind that the relationship between weight percent and moles EO is nonlinear, as shown in Fig. 3.17. The mathematical relationship between the two is as follows:

$$\text{wt \% EO} = \frac{(\text{moles EO}) \, (44)}{(\text{moles EO}) \, (44) + (\text{mol wt of alcohol})} \times 100$$

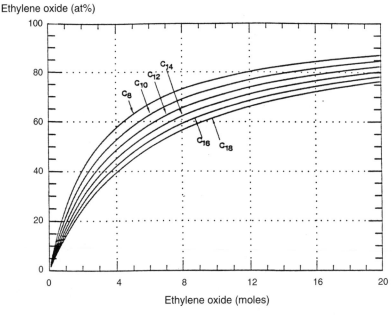

Figure 3.17   Relationship between wt % EO and moles EO for C_8 through C_{18} alcohol ethoxylates.

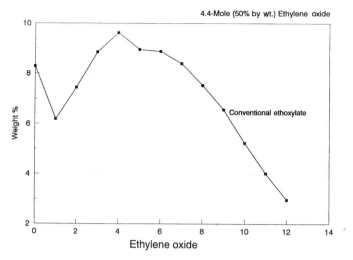

Figure 3.18 Ethoxymer distribution of conventional 4.4-mole dodecyl alcohol ethoxylate.

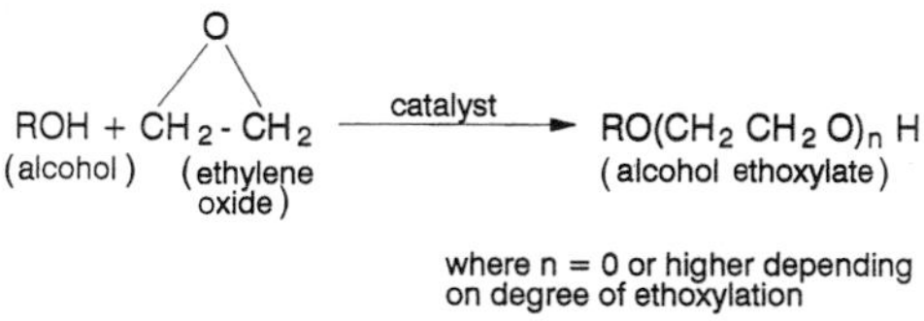

$$ROH + CH_2\text{-}CH_2 \xrightarrow{\text{catalyst}} RO(CH_2\,CH_2\,O)_n\,H$$

(alcohol) (ethylene oxide)   (alcohol ethoxylate)

where n = 0 or higher depending on degree of ethoxylation

Figure 3.19 Preparation of alcohol ethoxylate.

Product terminology normally describes ethoxylates in terms of their average weight percent EO or the average number of moles of EO. In reality, ethoxylates consist of a broad mixture of molecules (Fig. 3.18). During ethoxylation, a portion of the alcohol remains unethoxylated (remains as free alcohol), another portion ends up with one unit of EO attached, another with 2 units, and so on, thus yielding a distribution of ethoxymers or homologs. The distribution dips at the 1-mole ethoxymer because it is slightly more reactive to EO than both the unethoxylated alcohol and the higher mole ethoxymers [25].

Ethoxylation involves the addition of ethylene oxide one unit at a time to the terminal hydroxyl group of the molecule (Fig. 3.19). Conventional alcohol ethoxylates are catalyzed by either sodium hydroxide or potassium hydroxide. Peaked ethoxylates (see below) employ complex proprietary catalysts, which improve the reactivity between the alcohol or ethoxymer with EO. The result is a more "peaked" or "narrow range" distribution of ethoxymers. Figure 3.20 gives typical distributions for both conventional and "peaked" ethoxylates made from dodecanol. As shown, peaked ethoxylates typically contain less unethoxylated alcohol, less of the low and high mole ethoxymers, and more of the intermediate ethoxymers. The effects of distribution on performance is discussed below.

Regardless of whether an ethoxylate is conventional or peaked, increasing EO chain length increases solubility, solution viscosity, and melting point. Most ethoxylates having less than 50% EO are sulfated (see Section 3.3.3). Most water-soluble ethoxylates contain between 60 and 70% EO (approximately 7–9 moles EO).

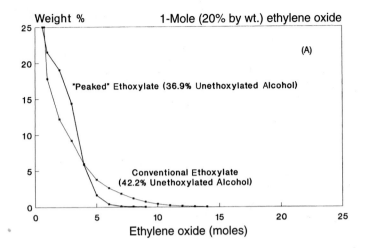

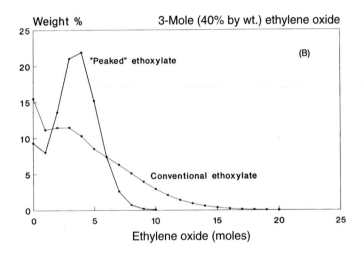

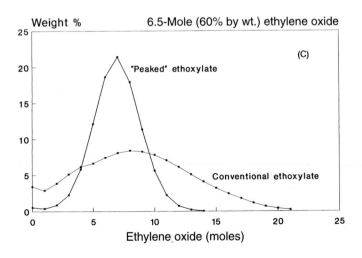

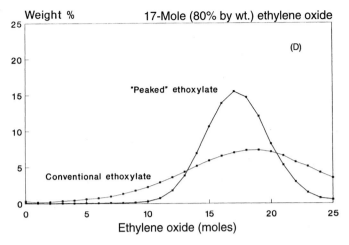

Figure 3.20   Ethoxymer distribution for conventional and peaked alcohol ethoxylates containing a) an average of 1-mole EO, b) an average of 3-moles EO, c) an average of 6.5-moles, and d) an average of 17-moles EO.

3.4.1.3   Properties and Characteristics

Physical Properties   Water solubility relates directly to EO chain length. Increasing EO content increases water solubility by increasing the number of sites along the EO chain that can associate with water molecules. Peaking the EO distribution also improves solubility by decreasing the concentration of the less-soluble, low mole ethoxymers.

Alcohol ethoxylates tend to form viscous gels at concentrations above about 25%. Increasing the alkyl carbon chain length and decreasing the EO chain length reduce the concentration at which gelling occurs. Gelling can be minimized by increasing EO content and by peaking the EO distribution.

Alcohol ethoxylates generally require heated storage. Ethoxylates having 60–70% EO generally melt in the 16–27 °C range (increasing EO increases melting point). Peaking the EO distribution lowers the melting point by about 2–5 °C because it effectively reduces the concentrations of the high mole ethoxymers.

Ethoxylates contain free alcohol, which can result in alcohol odor problems. A lower EO content increases alcohol odor by increasing free alcohol content. Shorter alkyl chain length increases odor because the alcohol itself is more pungent and volatile. Increasing alkyl carbon chain length, increasing EO content, and peaking the EO distribution all help to reduce alcohol odor.

Formulation in Powders   Alcohol ethoxylates are somewhat hygroscopic and are more difficult to formulate into powders than LAS and alcohol sulfates.

Fabric Detergency   Alcohol ethoxylates are excellent detergents, are known to excel at removal of liquid oily soil, and are not significantly affected by water hardness. The effects of alkyl carbon chain length and EO content have been well studied and documented [26,27]. For most applications, a lauryl-range alcohol with 60–70% EO (7–10 moles) ethoxylate is considered to be best. Optimum EO content, however, depends on use conditions, soil, and fabric.

Wetting   Alcohol ethoxylates are good wetting agents. An intermediate level of EO (between 50 and 60%) is generally best. A shorter alkyl chain length (octanol, decanol) is considered to be best at high surfactant concentrations, while a longer alkyl chain length (dodecanol, tetradecanol) is considered to be best at low surfactant concentrations. Peaking the EO distribution has also been shown to aid wetting.

Foaming   Alcohol ethoxylates produce a moderately stable foam (are considered to be low to moderate foamers). Optimum foam stability is obtained with a C_{10} or C_{12} alcohol with about 60–80% EO.

Alcohol ethoxylates are sometimes added to moderate the foaming of anionic surfactants. Although they reduce the amount of foam anionic surfactants can produce, they also stabilize the foam, reducing the effect on it of soil, mechanical action, and so on.

Compatibility with Additives   Alcohol ethoxylates are generally compatible with detergent additives (enzymes, quaternaries, perborate bleach). They are not compatible, however, with hypochlorite bleach (because the bleach reacts with the ethylene oxide chain) and the terminal hydroxyl group.

Mildness   Alcohol ethoxylate mildness is comparable to that of most anionic surfactants [20]. Although you might assume that since alcohol ethoxylates excel at oily soil removal, they would naturally defat skin when used in skin contact applications. Alcohol ethoxylates, however, have a low critical micelle concentration (cmc), which means that in solution these compounds exist predominantly in the form of micelles. This limits monomer concentration, which in turn limits the concentration of free surfactant available to defat skin.

Alcohol ethoxylate mildness is also dependent on EO chain length. A shorter EO chain length increases irritation potential.

3.4.1.4   Applications

Although alcohol ethoxylates are excellent detergents, they have not been incorporated into laundry powders to the same extent as LAS and alcohol sulfates because they contain free alcohol. Most powders are prepared by blending detergent components into a concentrated aqueous slurry (called a Crutcher slurry), which is then spray-dried to a powder (spraying the slurry into the top of a tower, where it meets a countercurrent stream of hot air). When ethoxylates are used, the free alcohol is volatilized during the spray-drying process, which condenses upon emerging from the spray tower, causing a cloud or plume. Pluming is undesirable from both aesthetic and regulatory points of view, since visible stack emissions are sometimes used to evaluate air quality. Peaked ethoxylates were initially developed to minimize free alcohol content, to make spray-drying an ethoxylate-based formulation easier [28]. In another approach being used to add ethoxylate to powders, the ethoxylate is sprayed onto spray-dried powder. This postdosing system works well for adding low levels of ethoxylate, but the primary surfactant must still be spray-dried. Agglomeration technology (the process of mixing dry ingredients, adding enough water to help produce a homogeneous mixture, and then removing the water through fluidized drying beds) continues to develop and will likely lead to an increase in the use of ethoxylate in detergent powders.

In liquid laundry detergents, alcohol ethoxylates are almost always used in combination with LAS, as either the principal or the secondary surfactant. Although a mixed active requires additional storage and more complex formulation, it does improve performance over a broader range of use conditions. High mole alcohol ethoxylates have also been shown to improve LAS performance in hard water by helping solubilize salts of calcium and magnesium LAS [19].

Alcohol ethoxylates are often used in hard-surface cleaners and a wide variety of industrial applications.

3.4.1.5 Availability

Alcohol ethoxylates are readily available as 100% active liquids or solids.

3.4.2   Alkylphenol Ethoxylates (APEs)

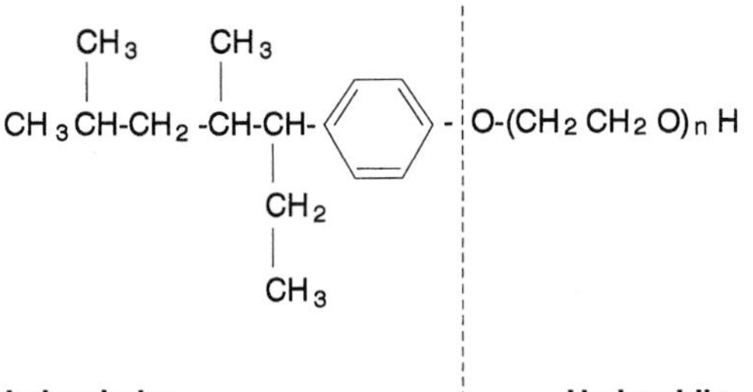

$$CH_3 CH-CH_2 -CH-CH- \langle\!\langle \bigcirc \rangle\!\rangle - O-(CH_2 CH_2 O)_n H \tag{10}$$

with CH_3 groups and CH_2—CH_3 branch

Hydrophobe
nonylphenol shown above (one of many possible isomers), octyl phenol, and dodecaphenol also produced

Hydrophile
(where n=1 to 100 depending on degree of ethoxylation)

Alkylphenol ethoxylates (APEs) are versatile surfactants that have been used for more than 40 years in a wide variety of applications. During the past decade, APEs have been criticized because they biodegrade more slowly than linear alcohol ethoxylates and because their biodegradation intermediates are more toxic to aquatic environments than the parent surfactant itself [29]. Recent risk analysis, however, indicates that APEs may not be accumulating in the environment and so may have been prematurely labeled an environmental concern [30].

APEs are used in a variety of industrial applications because they are easy to handle, available in a wide variety of ethoxylation levels, and sold in any volume from gallon jugs to tank trucks. They are about the lowest priced bulk surfactant on the market and are good surfactants. Approximately 1.4 billion pounds are annually consumed worldwide, including 500 million pounds in the United States.

3.4.2.1   Description of Hydrophobe

The hydrophobe of an alkylphenol ethoxylate is an alkylphenol. Most APEs are made from three feedstocks: octylphenol, nonylphenol, and dodecaphenol. Nonylphenol is made by linking three propylene groups together to form nonene (Fig. 3.21). Octene (for octaphenol) is produced by dimerizing isobutene, while dodecane (for dodecaphenol) is prepared by linking four units of propylene together. Increasing alkyl carbon chain length increases hydrophobicity, reduces water solubility, and results in more surface active molecules.

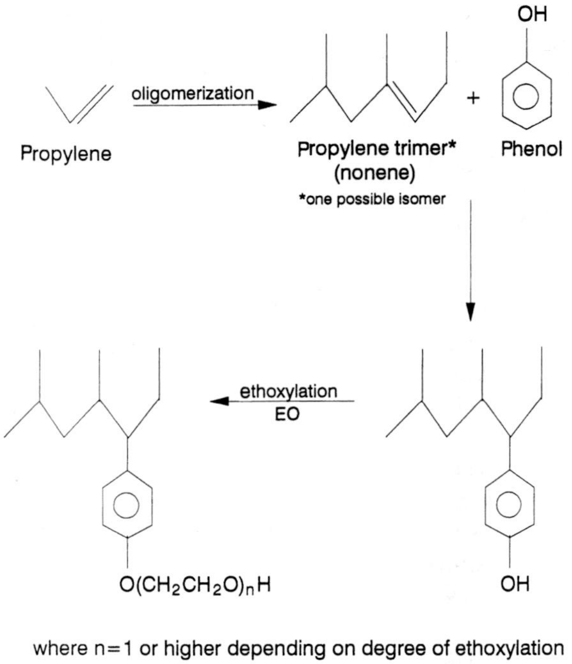

Figure 3.21   Preparation of nonylphenol ethoxylate (one of many possible isomers).

3.4.2.2   Description of Hydrophile

The hydrophile of APE is an ethylene oxide chain. Ethoxylation is accomplished in much the same way as with alcohol ethoxylates. The primary difference is that no unethoxylated alcohol (phenol) is obtained because the alkylphenol itself is more reactive than the ethoxymers.

Increasing EO chain length increases melting point, viscosity, and water solubility. Optimum surface properties, foaming, detergency, and wetting are obtained with an intermediate EO content (8–12 moles), which varies depending on the size of the hydrophobe.

3.4.2.3   Properties and Characteristics

Physical Properties   The presence of methyl branching in the alkyl chain increases solubility and decreases melting point, making alkylphenol ethoxylates, in general, easier to handle than their alcohol ethoxylate counterparts. APEs are easy to dissolve and offer less difficulty with gel formation than alcohol ethoxylates.

Formulation in Powders   Alkylphenol ethoxylates, like alcohol ethoxylates, are some-what hygroscopic and are more difficult than anionic surfactants to formulate in powders.

Fabric Detergency   APEs are good detergents. Optimum detergency is normally achieved with nonylphenol ethoxylate in the range of 8–12 moles of EO.

Wetting   Alkylphenol ethoxylates are good wetting agents. The general effects of EO content and carbon chain length on wetting are the same as with alcohol ethoxylates.

Foaming   Similar to alcohol ethoxylates.

Compatibility with Additives   Similar to alcohol ethoxylates.

Mildness   Alkylphenol ethoxylates are about as mild as alcohol ethoxylates and most anionic surfactants.

3.4.2.4   Applications

Alkylphenol ethoxylates are used in a wide variety of industrial applications. Their use in household products is limited because most major manufacturers avoid them owing to concern over APE biodegradation and aquatic toxicity, and because other surfactants are readily available. On the other hand, APEs are the largest volume nonionic surfactant used in industrial applications.

3.4.2.5   Availability

Alkylphenol ethoxylates are readily available as 100% active liquids.

3.4.3   Ethylene Oxide/Propylene Oxide (EO/PO) Block Copolymers

$$\overset{CH_3}{\underset{|}{}}$$

$$H(O\text{-}CH_2\text{-}CH_2)_m\text{-}O\text{-}(CH_2\text{-}CH\text{-}O)_n\text{-}(CH_2\text{-}CH_2\text{-}O)_{m^1} H$$

Ethylene oxide chain	Propylene oxide chain	Ethylene oxide chain
Hydrophile	**Hydrophobe**	**Hydrophile**

where m and m^1 = 0 or higher, n = 1 or higher

- -

PO/EO or PO/EO/PO block copolymer   (11)

$$\overset{CH_3}{\underset{|}{}} \qquad \overset{CH_3}{\underset{|}{}}$$

$$H(O\text{-}CH\text{-}CH_2)_m\text{-}O\text{-}(CH_2\text{-}CH_2\text{-}O)_n\text{-}(CH_2\text{-}CH\text{-}O)_{m^1} H$$

Propylene oxide chain	Ethylene oxide chain	Propylene oxide chain
Hydrophobe	**Hydrophile**	**Hydrophobe**

where m and m^1 = 0 or higher, n = 1 or higher

In comparison to other surfactants described in this chapter, EO/PO block copolymers do not represent a class of large-volume commodity surfactants, but they are important because they fit several important niches in the marketplace. EO/PO block copolymers are low foaming, nonionic surfactants that are mild to the skin and are good thickening and gelling agents.

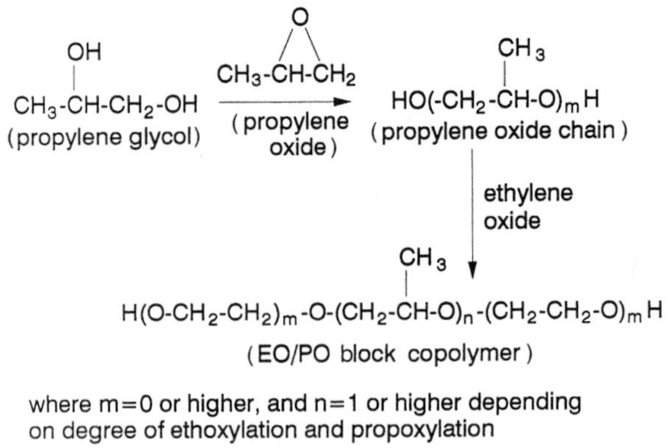

$$\text{(propylene glycol)} \quad CH_3\text{-CH-CH}_2\text{-OH} \xrightarrow[\substack{\text{(propylene}\\ \text{oxide)}}]{\substack{O\\ CH_3\text{-CH-CH}_2}} \quad HO(\text{-CH}_2\text{-CH-O})_m H \quad \text{(propylene oxide chain)}$$

ethylene oxide ↓

$$H(O\text{-CH}_2\text{-CH}_2)_m\text{-}O\text{-}(CH_2\text{-CH-O})_n\text{-}(CH_2\text{-CH}_2\text{-O})_m H$$

(EO/PO block copolymer)

where m=0 or higher, and n=1 or higher depending on degree of ethoxylation and propoxylation

Figure 3.22   Preparation of EO/PO block copolymer.

3.4.3.1   Description of Hydrophobe

The hydrophobe of an EO/PO block copolymer is the propylene oxide chain itself. This chain (Fig. 3.22) is made by reacting propylene glycol with a given amount of propylene oxide. The length of the polypropylene oxide chain influences the physical and surface chemical properties of the surfactant. In general, increasing molecular weight increases the ability of the surfactant to thicken various aqueous solutions and improves wetting properties. Decreasing chain length increases water solubility and up to a point gives better detergency.

3.4.3.2   Description of Hydrophile

The hydrophile of an EO/PO block copolymer consists of two poly(ethylene oxide) chains. As shown in Fig. 3.22, ethoxylation is accomplished by simply ethoxylating the propylene oxide chain. This gives a distribution of molecules composed of a block of propylene oxide with a block of ethylene oxide attached at each end (consequently it is called an EO/PO or, more properly, an EO/PO/EO block copolymer). The inverse molecule (i.e., PO/EO/PO) can be readily made by reversing the order of addition of the PO and EO. The reverse structures offer advantages in better defoaming and reduced tendency to gel. Increasing EO content improves the ability to thicken and to form gels and increases water solubility and detergency. Wetting properties of the surfactant are decreased, however.

3.4.3.3   Properties and Characteristics

EO/PO block copolymers are low-foaming surfactants used in automatic dishwashing detergents and in rinse aids. They are also used as emulsifiers, as solubilizers for flavors and fragrances, and as dispersants for for pigments and soils. In addition, they are used in cosmetic formulations because of their ability to thicken and gel, and because they are mild.

3.4.4   Alkyl Polyglycoside

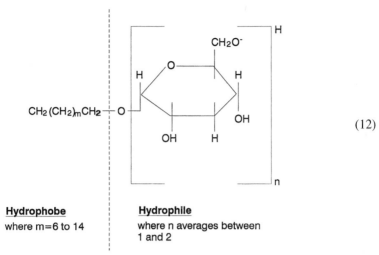

(12)

Hydrophobe	**Hydrophile**
where m=6 to 14	where n averages between 1 and 2

Alkyl polyglycosides were introduced in the early 1970s by Rohm & Haas as an ionic strength tolerant surfactant for industrial applications. The company later produced a purer, lighter colored material for cosmetic applications, then sold both products to Union Carbide. In the 1980s, A.E. Staley introduced a line of products to the detergent industry based on new process chemistry. To develop these products further, they formed a separate division called Horizon Chemical, which was sold to Henkel Corporation in 1988. Henkel began commercial production in 1992 from a 50 million pound per year plant located in Cincinnati, Ohio, which supplies product for the United States and Europe.

Alkyl polyglycosides are premium-priced, high-foaming nonionic surfactants that are mild and ionic strength tolerant. Although they have yet to capture significant market share, they are finding niche markets and do represent the first surfactant to employ an oleochemical feedstock as the hydrophilic portion of the surfactant molecule.

3.4.4.1   Description of Hydrophobe

The hydrophobe of an alkyl polyglycoside is usually a linear primary alcohol. Carbon chain length varies between C_8 and C_{16} depending on the feedstock (oleochemical or petrochemical) or on the desired solubility characteristics. Increasing alkyl carbon chain length increases surface activity but decreases solubility and salt tolerance. Foam stability also decreases with increasing alkyl carbon chain length.

3.4.4.2   Description of Hydrophile

The hydrophile of an alkyl polyglycoside is a polyglucose chain. Chain length can vary between one and more than eight units of glucose, but the distribution is highly skewed toward the 1- and 2-mole homologs. Process limitations currently make it impractical to achieve an average glucose chain length of below 1.4 or above 1.7 (Fig. 3.23). This somewhat narrow range, however, is broad enough to permit the adjustment of the hydrophilic behavior of the surfactant.

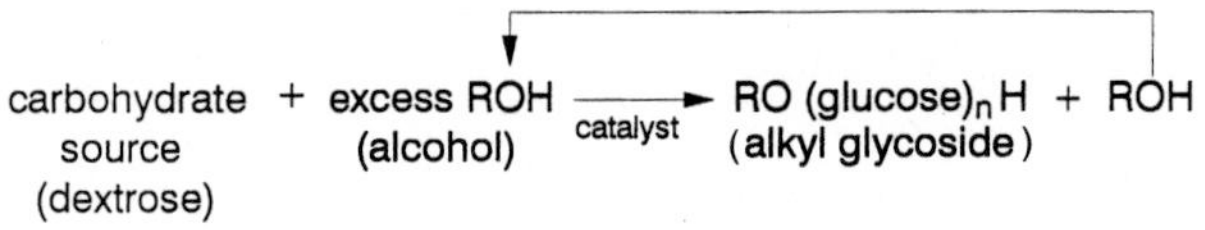

$$\text{carbohydrate} + \text{excess ROH} \xrightarrow{\text{catalyst}} \text{RO (glucose)}_n\text{H} + \text{ROH}$$

carbohydrate + excess ROH ——catalyst——▶ RO (glucose)$_n$H + ROH
source   (alcohol)   (alkyl glycoside)
(dextrose)

(where n averages between 1 and 2)

Figure 3.23   Preparation of alkyl polyglycoside.

Alkyl polyglycosides are manufactured from carbohydrates (usually high-dextrose-containing feedstocks) using processes based on Fisher glycosylation [31]. Yields are relatively low, and process conditions are carefully controlled to avoid by-products that detract from color and performance.

3.4.4.3   Properties and Characteristics

Physical Properties   Alkyl polyglycosides are very water soluble and tolerant of high ionic strength (high salt level) and water hardness ions. Solubility increases with increasing average glycoside chain length and decreasing average alkyl carbon chain length. Surface activity follows the opposite trends. Alkyl polyglycosides are stable at all pH values between 5 and 13.

Formulation in Powders   Little has been published regarding formulation of alkyl polyglycosides into powder or solid formulations. They have been found, however, to be useful as viscosity modifiers in Crutcher slurries.

Fabric Detergency   Alkyl glycosides do not perform as well as alcohol ethoxylates. Optimum detergency performance is normally obtained with a higher molecular weight alcohol and a lower level of polyglycoside.

Wetting   Increasing alkyl carbon chain length, and to a lesser extent decreasing glycoside chain length, improves surface properties and generally gives better wetting.

Foaming   Alkyl polyglycosides are better foamers than alcohol ethyloxylates, but they do not foam as well as LAS, AS, and AES. They are reported, however, to boost the foam of anionic surfactants [32].

Compatibility with Additives   Alkyl polyglycosides, like other nonionics, are compatible with cationic surfactants but not with hypochlorite bleach.

Mildness   Alkyl polyglycosides are reported to be mild and to increase the mildness of other surfactants found in solution [32].

3.4.4.4   Applications

Alkyl polyglycosides are relatively new to the detergent market and are being recommended for a variety of applications, including hand dishwashing detergents (because they are mild), spray-dried laundry detergents (because they help minimize water by lowering Crutcher viscosity), and liquid cleaners requiring ionic strength tolerance [32].

3.4.4.5   Availability

Alkyl polyglycosides are generally available as 50–70% aqueous solutions.

3.5   Nitrogen-Based Surfactants

Nitrogen-based surfactants are derived from fatty acids, olefins, and alcohols. This class of surfactants is broad and diverse because amines can be oxidized, alkylated, alkoxylated, quaternized, or otherwise derivatized to form a wide variety of surfactants. Approximately 900 million pounds of various amines are used worldwide each year. Two-thirds of this amount is used to make various quaternary surfactants for fabric and hair conditioning; the majority of the rest is oxidized or ethoxylated to make various specialty surfactants. Nitrogen-based surfactants can be cationic (quaternary ammonium salts), non-ionic (alkanolamides, ethoxylated amines), or amphoteric (amine oxides, alkylbetaines, amino acids).

3.5.1   Description of Hydrophobe

The hydrophobe of nitrogen-based surfactants consists of one or more alkyl groups derived from fatty acids or fatty alcohols (Fig. 3.24). Fatty acids can originate from oleochemical oils or animal fats, and alcohols can come from both oleochemical and petrochemical sources. Alkyl carbon chain length varies between C_8 and C_{18} depending on the feedstock and the composition desired.

3.5.2   Description of Hydrophile

The hydrophile of nitrogen-based surfactants is one or more nitrogen atoms associated with or without one or more oxygen containing groups. Hydrophilic character (and water

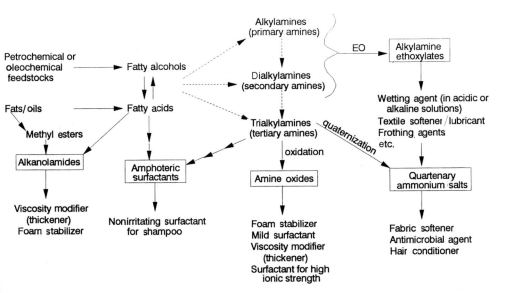

Figure 3.24   Preparation of various nitrogen-based surfactants.

solubility) of nitrogen-based surfactants can be increased by (a) attaching oxygen in the form of poly(ethylene oxide) chains, an oxide group, or hydroxyl groups to the nitrogen, (b) forming the amine salt by reacting the amine with an organic or inorganic acid, and (c) forming the quaternary ammonium salt by alkylating the nitrogen.

Nitrogen-based surfactants start with an alkyl amine or an amide (Fig. 3.24). Although primary and secondary amines can be used as is, most are ethoxylated or used to make tertiary amines, which in turn are used to make amine oxides, amphoterics, or quaternaries. Primary amines are also converted to polyamines, which are more surface active. Amides are usually produced as the alkanolamides. Each of these subclasses is discussed below.

3.5.2.1   Alkanolamides

$$
\begin{array}{cc}
O & X \\
\| & | \\
R\text{-}C\text{-}N\text{-}CH_2CH_2OH
\end{array}
$$

(13)

Hydrophobe	Hydrophile
where R=$C_{11\text{-}17}$ saturated or unsaturated alkyl carbon chain	where X=H for mono ethanolamide or -CH_2CH_2OH for diethanolamide

Alkanolamides are made by reacting fatty acids or fatty methyl esters with ethanolamine or diethanolamine. Methyl esters result in high active alkanolamides called "superamides." Alkanolamides can also be produced by ethoxylating fatty amides.

Alkanolamides are nonionic surfactants, since they do not ionize in solution. Surfactant structure can vary in the number of ethanol groups attached to the nitrogen, and in carbon chain length of the alkyl chain.

Alkanolamides are considered to be excellent cosurfactants. They affect the physical and surface chemical properties of other surfactants in solution via hydrogen bonding. This interaction leads to increased viscosity (thickening), and increased foam elasticity (foam stabilization). By themselves, alkanolamides are also used as antistatic agents, as anticorrosion agents, and as wetting agents.

3.5.2.2   Amine Ethoxylates

$$
\begin{array}{c}
X \\
| \\
R\text{-}N\text{-}(CH_2CH_2\text{-}O)_n H \\
| \\
Y
\end{array}
$$

(14)

Hydrophobe	Hydrophile
where R=$C_{12\text{-}18}$ alkyl carbon chain	where X and Y are alkyl groups or other ethoxylate chains

Primary and secondary amines can be ethoxylated to increase their solubility. Although a wide variety of amines are ethoxylated, the general performance characteristics of ethoxylated amines are similar. They are generally soluble over a large pH range and show good emulsification and wetting properties in both acidic and alkaline solutions. Consequently, ethoxylated amines are used in the textile industry as dye leveling agents (to aid in distributing the dye) and lubricants, and as emulsification agents for asphalt.

3.5.2.3   Quaternary Ammonium Salts

$$
\left[\begin{array}{c} R \diagdown \qquad \diagup R' \\ N^+ \\ R'' \diagup \qquad \diagdown R''' \end{array} \right] X^- \qquad \text{where } X^- = Cl^-, \tfrac{1}{2}SO_4^{-2}, \text{ etc.}
$$

(15)

Hydrophobe
R, R', R'', R''' = alkyl or alkyl aryl
groups (generally 2 or 3 are methyl groups)

Hydrophile

$$
\diagdown \qquad \diagup \\ N^+ \\ \diagup \qquad \diagdown
$$

Any tertiary amine can be alkylated or quaternarized (i.e., have a fourth group attached to the nitrogen) to form a positively charge nitrogen, called a quaternary ammonium salt. Quaternization can be accomplished with dimethyl sulfate, benzyl chloride, ethylene oxide, or an alkyl halide. The parent amine can be simple trialkylamine, a substituted imidazoline, or an alkoxylated amidoamine.

Quaternary ammonium salts (usually called quats) are cationic surfactants that carry a net positive charge in solution. They adsorb strongly onto negatively charged surfaces. Adsorption of cationics on surfaces modifies their properties by:

- Making surfaces more greasy (feel "slicker"), or lipophilic (a property used for improving the adhesion of asphalt onto aggregates)
- Helping to reduce static charge buildup (key property for hair and fabric softeners)
- Interfering with microbial activity (biocides, sanitizers)

Quats are soluble in both acidic and alkaline aqueous solutions. Increasing alkyl carbon chain length, the number of long alkyl carbon chains, and the degree of saturation increases hydrophobicity. Some quats are available as the ammonium chloride salt, others are available as the methosulfate salt (OSO_2CH_3), depending on the quaternization agent used. Methosulfate salts are generally more thermally stable than their corresponding halide salts and are less corrosive to metal.

Hair conditioners generally use quats of the alkyl ($C_{12}-C_{18}$) trimethyl or dialkyldimethyl type. Fabric softeners for dryer sheets generally use the methosulfate salt of a dialkyl ($C_{16}-C_{18}$)dimethyl type of quat, while more soluble salts (unsaturated dialkyldimethyl quats, amido imidazolinium quats, diamido alkoxylated quats) are used in combination with dialkyldimethyl-type quats in liquid softeners designed to be added to laundry during the rinse cycle. Antimicrobial agents generally use the quats of the alkyl($C_{12}-C_{14}$)benzyldimethyl type.

3.5.2.4   Amine Oxides

$$CH_3$$
$$R-N \longrightarrow O \quad (16)$$
$$CH_3$$

Hydrophobe   **Hydrophile**
where R = saturated
or unsaturated C_{12}-C_{18}
alkyl chain

Tertiary amines, normally alkyldimethylamine or alkyldiethanolamine, can be oxidized with hydrogen peroxide to produce amine oxides. Amine oxides are unique in that they act as nonionics in neutral or alkaline solutions, but as cationics in acidic solutions (i.e., will precipitate with anionic surfactants). Like alkanolamides, amine oxides interact with other surfactants through hydrogen bonding and are therefore effective in altering their physical and surface chemical properties. Generally, a C_{12}–C_{18} alkyldimethylamine will be used as the parent amine.

The major application for amine oxides is as foam stabilizers (also called "foam boosters"), used principally in hand dishwashing liquids containing alcohol sulfate and alcohol ether sulfate. Amine oxides are also used in a variety of applications where solubility under high ionic strength conditions is important and when compatibility with strong oxidants (hydrogen peroxide, sodium hypochlorite) is important. They are also mild (when used at low concentrations) and, like alkanolamides, are good thickening agents.

3.5.2.5   Amphoteric Surfactants

$$CH_3 \quad O$$
$$R-N^+-CH_2C-O^- \quad (17)$$
$$CH_3$$

Hydrophobe   **Hydrophile**
where R = alkyl or
alkyl/aryl chain

Surfactants containing both an anionic (negatively charged) and a cationic (positively charged) functional group are called amphoteric surfactants. The positive charge is provided by a quaternary nitrogen, and the negative charge is normally supplied by a fatty acid carboxylate group.

The most common type of amphoteric is an alkylbetaine. As shown in Fig. 3.25, alkylbetaines are produced by reacting alkyldimethylamine with sodium chloroacetate. Other betaines, such as amidoalkylbetaines and imidazolinium-type betaines, can be produced from fatty acids. A large variety of amphoterics is available, including those that contain sulfonate groups as the anionic functionality (sulfobetaines or sultaines).

Amphoterics are relatively expensive surfactants that are used when compatibility with other ingredients, or compatibility with skin, is important. They are used in baby shampoos and as cosurfactants in hand dishwashing liquids. Amphoterics are also used where water hardness, ionic strength, and pH prevents other surfactants from being used. In fact, the water solubility of betaines actually increases with ionic strength.

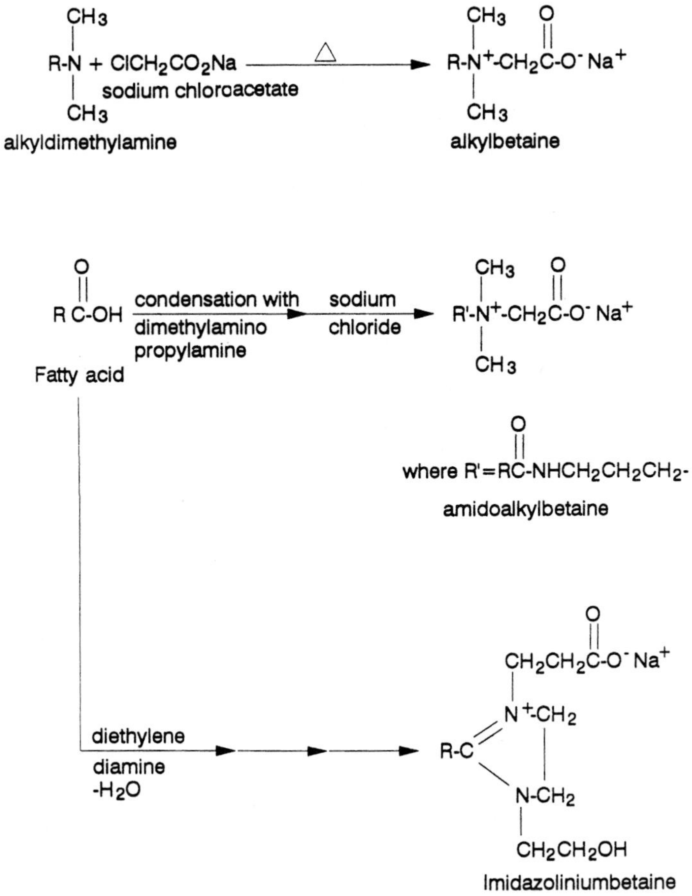

Figure 3.25   Preparation of various betaines.

3.5.2.6   *Polyamines*

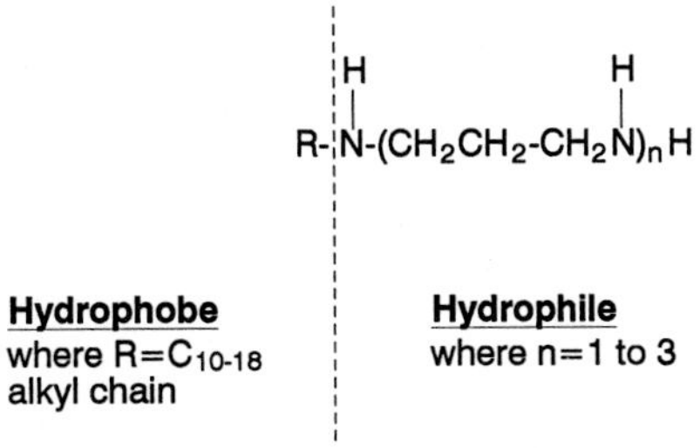

$$R-\overset{\overset{\displaystyle H}{|}}{N}-(CH_2CH_2-CH_2\overset{\overset{\displaystyle H}{|}}{N})_n H$$

(18)

Hydrophobe
where $R=C_{10-18}$
alkyl chain

Hydrophile
where $n=1$ to 3

Primary monoamines can be converted to diamines by reaction with acrylonitrile (CH_2 CH—CN) followed by hydrogenation. The diamine can again be reacted with acrylonitrile, followed by hydrogenation to make the triamine.

Polyamines are more surface active than their corresponding monoamines and are used in applications in metal treating and pigment dispersing.

3.6   Getting More Information

As mentioned earlier, being well-informed makes selecting the right surfactant less difficult, and becoming well-informed is not difficult. There are a number of good sources of readily available, up-to-date information.

3.6.1   Surfactant Suppliers

Most surfactant suppliers spend a considerable amount of effort gaining expertise on the use of their surfactants (and those of their competitors!) in a variety of applications. It is important to the suppliers from a commercial standpoint to help you use their surfactants. It is also important to them to make sure you can use their surfactants in your application. They accomplish this by filing patents or making public disclosures, all of which takes applications research. Many suppliers have information already prepared to meet your needs, or can easily access it from their archives. Some will also perform lab tests to help you develop the right surfactant for your application. All you have to do is ask. You do not have to divulge proprietary details of your applications, although the more detail you give, the better the supplier will be able to meet your needs. Keep in mind, however, that the supplier's goal is to determine which surfactant in his product line best fits your needs. It is up to you to know what alternative surfactants are available.

The best source on who makes what surfactants is *McCutcheon's* [5].

3.6.2   Industry Associations

The Soap and Detergent Association (SDA) is an industry group composed of permanent staff and representatives from both surfactant manufacturers (Shell, Vista, Monsanto, etc.) and surfactant users (Procter & Gamble, Lever, Colgate, etc.). The SDA addresses major legislative and regulatory issues affecting the surfactant and detergent industry, and is a good source of information on these types of issues. For more information, contact the SDA at 475 Park Avenue South, New York, New York 10016 (telephone: 212-725-1262).

The Chemical Specialty Manufacturers Association (CSMA) is similar to the SDA but addresses the needs of a broader membership and tackles a greater variety of tasks, including general education of the industry, test method development, and specific legislative issues. For more information, contact the CSMA at 1930 Eye Street, NW, Washington, DC 20006.

Committee D12 of the American Society for Testing and Materials (ASTM) is active in developing test methods and guidelines for evaluating surfactant performance. For more information, contact the ASTM at 1916 Race Street, Philadelphia, Pennsylvania 19103.

3.6.3   Surfactant Publications

The *Journal of the American Oil Chemists' Society (JAOCS)* devotes two or three issues per year strictly to surfactants and surfactant applications. *Tenside* is also a good source of practical information, although it is somewhat expensive. The *Journal of Colloid and Interface Science* is a good means of learning what's new in fundamental surface science.

Publications that contain applications-oriented, general-interest articles (usually written by suppliers or consultants) are *HAPPI* (Household and Personal Products Industry) and *Soap/Cosmetics/Chemical Specialties*.

For more information, contact:

- *JAOCS*   American Oil Chemists' Society, P.O. Box 3489, Champaign, IL 61826-3489 (217-359-2344)
- *Tenside*   Hanser Publishers, Kolbergerstr. 22, Postfach 86 04 20 D-8000, Munich 86, Germany
- *Journal of Colloid and Interface Science*   Academic Press, Inc., 6277 Sea Harbor Drive, Orlando, FL 32887-4900
- *HAPPI*   Rodman Publishing Corporation, Box 555, 17 South Franklin Turnpike, Ramsey, NJ 07446 (201-825-2552)
- *Soap/Cosmetics/Chemical Specialties*   PTN Publishing Company, 445 Broad Hollow Road, Melville, NY 11747 (516-845-2700)

3.6.4   Annual Meetings/Short Courses

The annual meeting of the American Oil Chemists' Society (Surfactants and Detergents Division) is a good source of information on what's new in the surfactants and detergents industry. Each meeting usually consists of six half-day sessions covering applications and ingredient-oriented material, with a session or two of basic surface chemistry. If more fundamental surface science is desired, attend the national meeting of the Surface and Colloid Science section of the American Chemical Society.

References

1. Rosen, M.J., *Surfactants and Interfacial Phenomena*, Wiley, New York, 1978, p. 162.
2. Schwarz, E.G., Reid, W.G., Ind. Eng. Chem., *56*, 26 (1964).
3. Cox, M.F., Matson, T.P., J. Am. Oil Chem. Soc., *61*, 1273, (1984).
4. Ferguson, J., Proc. R. Soc., London, *127B*, 387 (1939).
5. *McCutheon's Emulsifiers and Detergents 1993 North American Edition*, McCutcheon Division, MC Publishing, Glenrock, NJ, 1993.
6. Griffin, W.C., J. Soc. Cosmet. Chem., *1*, 311 (1949).
7. Griffin, W.C., J. Soc. Cosmet. Chem., *5*, 249 (1954).
8. Davies, J.T., Proc., 2nd Int. Congr. Surface Activity, London, *1*, 426 (1957).
9. Becher, P., *Emulsions: Theory and Practice*, 2nd ed., ACS Monographs Ser. No. 162, Reinhold, New York, 1965, p. 249.
10. Brandrup, J., Immergut, E.H., Eds., *Polymer Handbook*, Wiley, New York, 1966, p. IV-341.
11. Shinoda, K., Arai, H., J. Phys. Chem., *68*, 3485 (1964).
12. Shinoda, K., Arai, H., J. Colloid. Sci., *20*, 93 (1965).
13. Morris, P.A., Wharry, D.L., Roheim, J.R., U.S. Patent 5,145,595.
14. Tjepkema, J.J., Paulis, B., Huijser, H.J., in *Proceedings: Chemicals from Petroleum and Natural Gas*, Vol. IV, Fifth World Petroleum Congress, New York, 1959, p. 237.
15. Nielson, A.M., Britton, L.N., Russell, G.L., McCormick, T.P., Filler, P.A., Microbial mineralization of dialkyltetralinsulfonate (DATS) in soil and aquatic systems, *Environ. Sci. Technol*, 1993 (submitted for publication).
16. Putnik, C.F., McGuire, S.E., J. Am. Oil Chem. Soc., *55*, 909 (1978).
17. Cox, M.F., J. Am. Oil Chem. Soc., *66*, 1637 (1989).

18. Schwuger, M.J., in ACS Symposium Series No. 253, American Chemical Society, Washington, DC, 1984.
19. Cox, M.F., Matheson, K.L., J. Am. Oil Chem. Soc., *62,* 1396 (1985).
20. Singer, E.J., Pittz, E.P., in *Surfactants in Cosmetics*, M.M. Rieger, Ed., Dekker, New York, 1985, p. 133.
21. Smith, D.L., J. Am. Oil Chem. Soc., *68*, 629 (1991).
22. Cox, M.F., J. Am. Oil Chem. Soc., *67*, 599 (1990).
23. Satsuki, T. INFORM, *3*, 1099 (1992).
24. Drozd, J.C., in *World Conference on Oleochemicals into the 21st Century*, American Oil Chemists' Society, Champaign, IL, 1991, p. 256.
25. Schachat, N., Greenwald, H.L., in *Nonionic Surfactants*, M.J. Schick, Ed., Dekker, New York, 1966, p. 28.
26. Cox, M.F., J. Am. Oil Chem. Soc., *66*, 367 (1989).
27. Raney, K.H., J. Am. Oil Chem. Soc., *68*, 525 (1991).
28. Wharry, D.L., Sones, E.L., McGuire, S.E., McCrimlisk, G., Lovas, J., J. Am. Oil Chem. Soc., *63*, 691 (1986).
29. Marcomini, A., Filipuzzi, F., Giger, W., Chemosphere, *17*(5), 853 (1988).
30. Naylor, C.G., *Soap/Cosmet./Chem. Spec.*, Aug. 27, 1992.
31. Schulz, P., Chim. Oggi, *10*, 33 (1992).
32. Allen, P.S. in *Proceedings: CMRA Meeting, January 25, 1993, Cincinnati, OH*, to be published by Chemical Management and Resource Association, Staten Island, NY.

CHAPTER 4

Bleaches and Optical Brighteners

Olina G. Raney

Both peroxygen and hypochlorite bleaching agents are used in household laundry products. While hypochlorite bleaches are found in bleach additive formulations, various types of peroxygen bleach have been developed for detergent products as well as for bleach additives. These include bleach activators for peroxygen bleaches that generate peroxycarboxylic acids in the washing solution, preformed peroxycarboxylic acids, and metal-catalyzed peroxygen bleaching systems.

Various factors affect bleach performance, the formulation stability, and choice of the bleach system to be used. These factors are discussed, and data showing the performance of various bleach systems are provided.

The use of optical brighteners for enhancing the whiteness and brightness of household laundry is also discussed in this chapter. Optical brighteners of various types have been developed. These have affinity to different types of fibers, and some have better bleach stability. The requirements for an effective optical brightener are discussed, as well as the factors that affect their selection and performance.

4.1   Bleaches

4.1.1   Introduction

The surfactants and enzymes formulated into detergent products remove the oily and pro-teinaceous soils from household laundry. For the removal of stains such as tea, red wine, chocolate, or fruit juices, bleach is needed. Bleaches may be incorporated into the detergent formulation itself but are most often added as a separate product.

In general, bleaching is any color change of a stain in the direction of decolorization. The traditional bleachable soils are chromophoric systems with conjugated carbon double bonds in polymethine chains, or quinoidic systems that acquire the properties of a dye through the presence of amino, hydroxy, or carboxylic groups. Typical bleachable soils consist of natural substances such as brown tannins (tea, red wine), red to blue anthocyanins (cherry, black-berry, red currant), curcuma dyes (curry, mustard), organic polymers of the humic acid type (coffee, tea, cocoa), pyrrol derivatives (chlorophyll), and carotinoid dyes (carrots, tomatoes) (Fig. 4.1). Synthetic dyes used in cosmetics, hair colorants, and inks may also be present in household stains.

Bleaching agents irreversibly oxidize and decolorize the bleachable soils that are present on fabrics. Two types of bleaching agent, those containing peroxygen atoms and those

Olina G. Raney, 2602 Barrington Court, Sugar Land, Texas 77478, U.S.A.

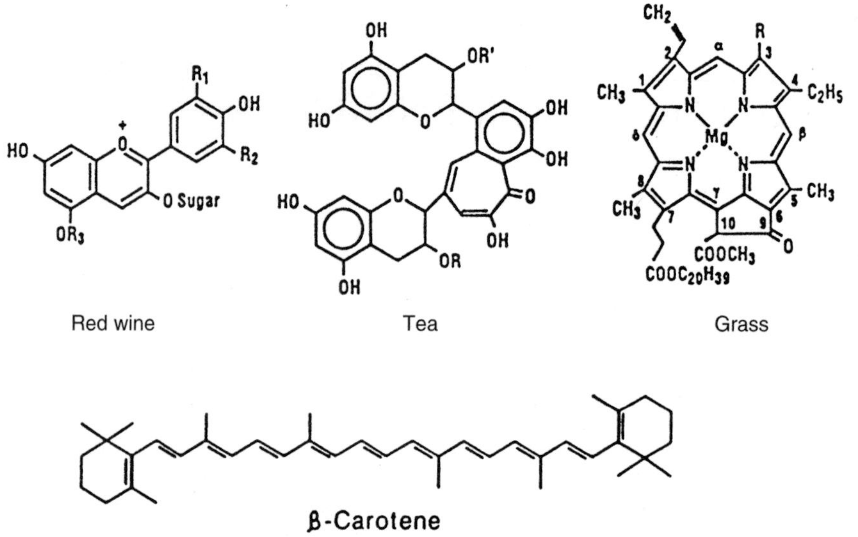

Figure 4.1   Bleachable soils from natural substances.

containing hypochlorite ions, are typically used in household laundry. While peroxygen bleaches have been predominantly used in Europe because of their effectiveness at high washing temperatures, hypochlorite bleaches have traditionally been preferred in the United States, where washing temperatures are lower. Recently, however, some powder detergent products in the United States have used peroxygen bleaches, and this trend is expected to continue.

Peroxygen bleaches are not stable in typical aqueous heavy-duty liquid formulations. Recent developments have led to the incorporation of peroxygens in nonaqueous heavy-duty liquid detergents. However, the discussion in this section focuses primarily on the use of bleaches in powder formulations.

4.1.2   Hypochlorite Bleaching Agents

Hypochlorite bleach is typically used as an additive to the detergent during the washing process. It is also used as a key ingredient in hard-surface cleaners and sanitizers.

Sodium hypochlorite is the most commonly used hypochlorite bleach in laundry processes [1]. Other sources of active chlorine are compounds such as potassium, lithium, or calcium hypochlorite, chlorinated trisodium phosphate, and sodium or potassium dichloroisocyanurate. The latter are generally not used in household laundry products but have uses in sanitizers and in automatic dishwashing formulations.

Sodium hypochlorite solutions are made by the reaction of liquid chlorine with a solution of sodium hydroxide. The strength of a hypochlorite bleach solution is expressed by its available chlorine content, which is determined by the electron transfer capacity of the oxidizing chemical. The hypochlorite ion accepts two electrons during its conversion to a chloride and, therefore, the available chlorine is calculated as twice the weight percent of chlorine in the hypochlorite molecule.

The pH of the hypochlorite laundry product is an important consideration in its formulation. As the acidity is increased, hypochlorite becomes more unstable owing to the formation of HOCl.

$$OCl^- + H^+ \rightleftharpoons HOCl \tag{4.1}$$

At room temperature, the pK_a of this equilibrium is 7.5. HOCl is more unstable than OCl^- While reducing the pH leads to more HOCl formation, which is desired for bleaching, it reduces the storage stability of the product. Therefore, most commercial products have a pH above 12. The stability of hypochlorite is also dependent on temperature and initial ionic concentration.

In general, the stain removal performance of hypochlorite bleach is superior to that of peroxygen bleaches. Also, it is effective as an antibacterial agent, even at room temperature. However, a major disadvantage of hypochlorite bleach is that it is too aggressive for several fabrics and dyes and often results in fabric damage and color fading.

Although hypochlorite bleaches are commonly used as bleach additives in household laundry in the United States, laundry detergents containing hypochlorite bleaches are not available. This is primarily due to the aggressive nature of the hypochlorite, which limits the chemical stability of the surfactants, optical brighteners, enzymes, and perfumes in the formulation.

4.1.3   Peroxygen Bleaching Agents

Peroxygen bleaching agents [1–4] are typically derived from *hydrogen peroxide*. While hydrogen peroxide is used in some liquid bleach formulations, its various derivatives are used in the powdered detergent and bleach formulations. The most commonly used are *sodium perborate tetrahydrate* ($NaBO_3 \cdot 4H_2O$) and *sodium perborate monohydrate* ($NaBO_3 \cdot H_2O$). As shown in Fig. 4.2, sodium perborate tetrahydrate is a peroxodiborate structurally. Thus, the name "sodium perborate tetrahydrate" is actually a misnomer since it was given based on the chemical composition of this compound and prior to the determination of the precise chemical structure. When dissolved in the wash liquor, hydrolysis of the anionic structure produces hydrogen peroxide, the bleaching agent. Sodium perborate tetrahydrate has been commonly used in household laundry, especially in Europe.

Recently developed U.S. detergent formulations have used sodium perborate monohydrate. This compound does not contain the level of hydrated water found in sodium perborate tetrahydrate and therefore has a higher active oxygen content (15% active oxygen content vs. 10% in the tetrahydrate). In comparison to the tetrahydrate, the monohydrate also has

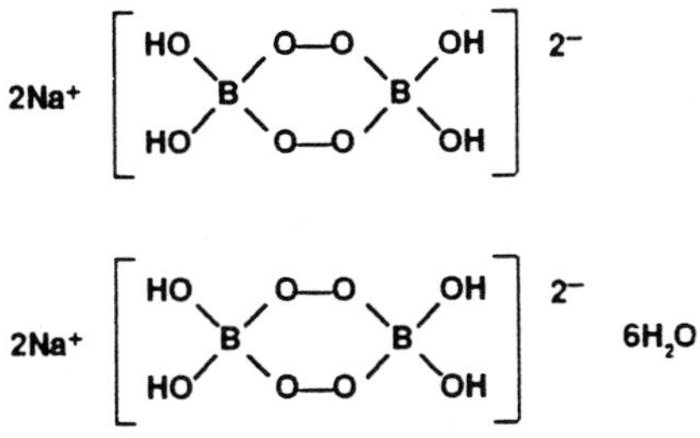

Figure 4.2   Sodium perborate and sodium perborate tetrahydrate.

improved stability, better compatibility with other detergent ingredients, and better low temperature dissolution characteristics.

Another peroxygen used is *sodium percarbonate* or sodium carbonate peroxyhydrate ($Na_2CO_3 \cdot 1.5H_2O_2$). This compound, in comparison to the perborates, contains oxygen in the form of a hydrogen peroxide adduct. It is not as stable as the perborates when in contact with other detergent ingredients and has, therefore, been used primarily in bleach additive formulations. However, recent advances in product stability have resulted in the incorporation of sodium percarbonate in detergent powder formulations in Japan [5].

The strength of various peroxygen compounds can be compared by using *active oxygen content* as a basis. All peroxygens contain the group –O–O– in which two oxygen atoms are linked together. The –O–O– link in these compounds can be broken, resulting in the liberation of one oxygen atom, termed *active oxygen*. The percentage of active oxygen can thus be calculated as the atomic weight of the "active oxygen" atoms present in the peroxygen compound divided by the molecular weight of the compound. For example, the active oxygen content of 100% w/w hydrogen peroxide is calculated as $(16 \div 34) \times 100 = 47.05\%$. For aqueous solutions, the concentration must be taken into account. Therefore, the active oxygen content of 35% strength hydrogen peroxide is calculated as $(47.05 \times 35) \div 100 = 16.47\%$.

The chemical reactions that occur during bleaching are mechanistically complex. This is due to the wide variety of bleachable soil structures present in household laundry. It is generally believe that bleaching results from the formation of the perhydroxyl anion, OOH^-, which is produced from the hydrogen peroxide source as follows:

$$H_2O_2 + OH^- \rightarrow H_2O + OOH^- \tag{4.2}$$

As the equation indicates, the concentration of the perhydroxyl anion is dependent on the pH of the system. Figure 4.3 shows the degree of dissociation of hydrogen peroxide and the generation of the perhydroxyl anion with increasing pH. The pK_a for hydrogen peroxide— that is, the pH at which half of the hydrogen peroxide is disassociated—is 11.7 at 20 °C.

Increasing pH, temperature, and bleaching agent concentration all increase the concentration of perhydroxyl anion in a washing solution and, therefore, improve stain removal.

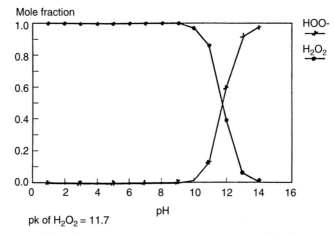

Figure 4.3   Degree of dissociation of hydrogen peroxide and generation of perhydroxyl anion HOO⁻ as a function of pH.

Other factors that determine bleaching effectiveness are wash time, interactions between fabric, soil, and bleach, and the presence of inorganic salts and complexes.

A reduction in bleaching effectiveness sometimes occurs as a result of depletion in solution, or in the formulation itself, of hydrogen peroxide and its derivatives by decomposition, as indicated by the following equation:

$$2H_2O_2 \rightarrow 2H_2O + O_2 \qquad (4.3)$$

The presence of metal ions such as copper, iron, and manganese can lead to the catalytic decomposition of the hydrogen peroxide. The effect of these catalysts can be minimized by the addition of stabilizers such as magnesium silicate.

4.1.4   Bleach Activators

Bleach activators are organic compounds that react with a peroxygen bleach to yield a *peroxycarboxylic acid*. This reaction, which is termed *perhydrolysis*, occurs in the wash liquor (Fig. 4.4). A distinct advantage of peroxycarboxylic acids is that they exhibit improved bleaching characteristics over hydrogen peroxide or perborates, even at low washing temperatures.

The most well-known bleach activator, and the one most commonly used in Europe, is *tetraacetylethylenediamine* (TAED) (Fig. 4.5). Only two of its four acetyl groups are available for perhydrolysis, yielding peracetic acid. In Europe, detergents containing TAED typically contain 12–24% sodium perborate tetrahydrate and 1–3% TAED.

Recent advances in activator technology have led to the development of systems that generate long-chain surface active peroxycarboxylic acids. These often exhibit performance superior to that of the shorter chain peroxycarboxylic acids such as peracetic acid. Activators in the form of acyloxybenzenesulfonates (Fig. 4.6) with chain lengths in the C_6–C_{10} range have been introduced in commercial detergent formulations in the United States [6].

Surface active peroxycarboxylic acids can also be generated using *p*-sulfophenyl carbonate activators [7], which perhydrolyze to form peroxycarboxylic acids containing

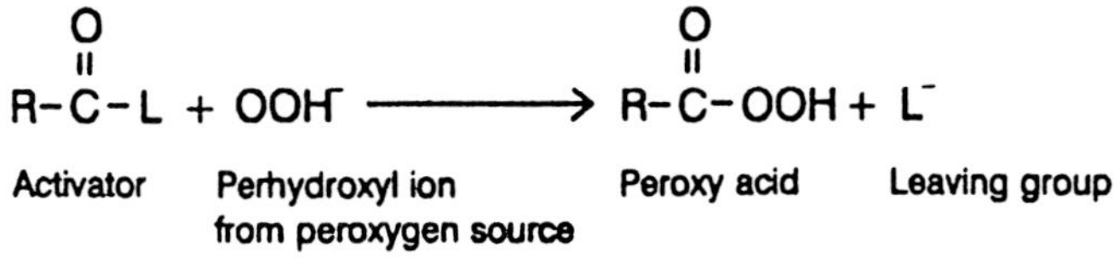

Figure 4.4   The perhydrolysis reaction.

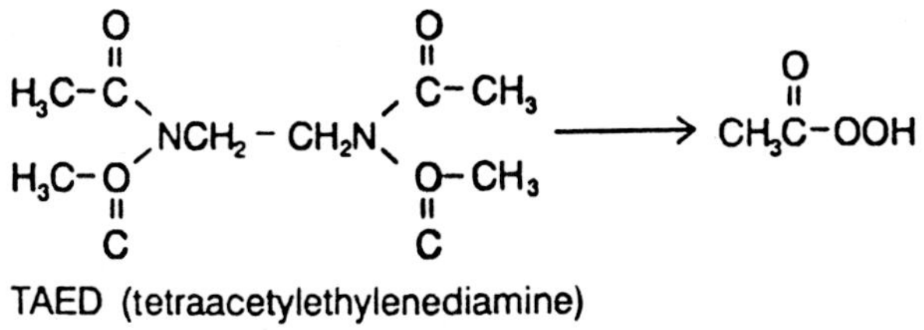

TAED (tetraacetylethylenediamine)

Figure 4.5   The perhydrolysis of the TAED molecule to yield peracetic acid.

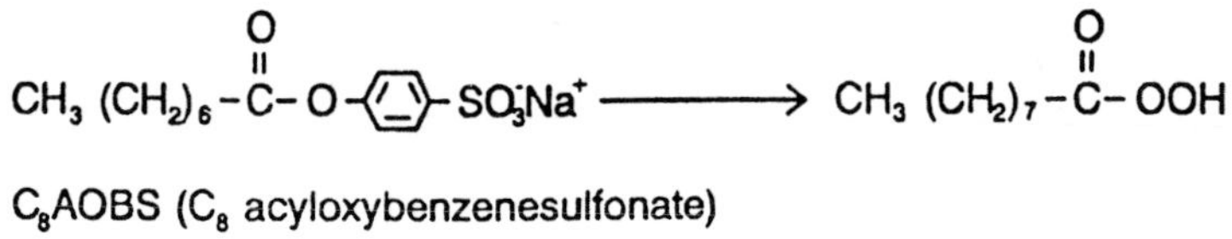

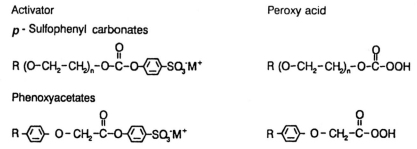

C$_8$AOBS (C$_8$ acyloxybenzenesulfonate)

Figure 4.6 / The perhydrolysis of the C$_8$ acyloxybenzenesulfonate molecule to yield peroxynonanoic acid.

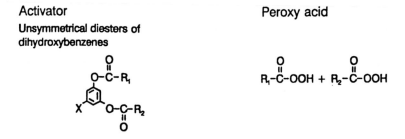

Figure 4.7   Use of activators to form pexoxy acids containing ethylene oxide groups.

Activator   **Peroxy acid**

**Unsymmetrical diesters of
dihydroxybenzenes**

Figure 4.8   Bleach activator capable of generating a mixed monoperoxycarboxylic acid.

ethylene oxide groups (Fig. 4.7). By varying the chain length and the number of ethylene oxide groups in the molecule, peroxycarboxylic acids with optimum hydrophobe–hydrophile characteristics can be generated. Phenoxyacetates [8] can also produce peroxycarboxylic acids with modified hydrophobe properties.

Bleach activators that generate mixed monoperoxycarboxylic acids have been developed (see, e.g., Fig. 4.8) [9]. Since a variety of bleachable soils is present in household laundry, overall bleaching is improved by generating peroxycarboxylic acids of different chain lengths. Activators that generate diperoxycarboxylic acids have also been described in the literature [10].

Also of interest are bleach activators that generate quaternary ammonium compounds upon perhydrolysis (Fig. 4.9) [11,12]. In this case, both the activator and the peroxycarboxylic acid generated are substantive to the fabric surface, thus resulting in improved bleaching performance. These compounds provide fabric softening in addition to bleaching.

The factors affecting the performance of bleach activator–peroxygen systems include the pH of the system, the chain length of the peroxycarboxylic acid generated, and the ratio of activator to peroxygen.

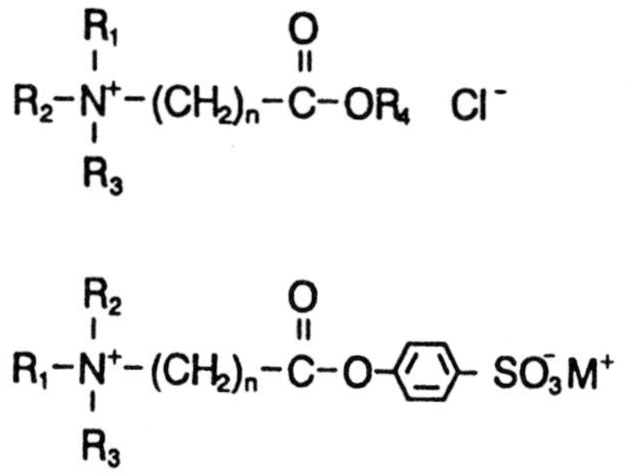

$$R_2-\overset{\overset{\displaystyle R_1}{|}}{\underset{\underset{\displaystyle R_3}{|}}{N^+}}-(CH_2)_n-\overset{\overset{\displaystyle O}{\|}}{C}-OR_4 \quad Cl^-$$

$$R_1-\overset{\overset{\displaystyle R_2}{|}}{\underset{\underset{\displaystyle R_3}{|}}{N^+}}-(CH_2)_n-\overset{\overset{\displaystyle O}{\|}}{C}-O-\langle\underline{\quad}\rangle-SO_3^-M^+$$

Figure 4.9   Bleach activators that can be perhydrolyzed to yield quaternary ammonium compounds.

The optimum bleaching performance for peroxycarboxylic acids is in the pH range 8–9, corresponding to their pK_a value. However, a higher pH is required for optimum detergent performance and for complete perhydrolysis of the bleach activator, leading to generation of peroxycarboxylic acid.

The optimum chain length for the monoperoxycarboxylic acid is C_8–C_{10}, while the optimum chain length for diperoxycarboxylic acid is about C_{12}.

The optimum *activator-to-peroxygen mole ratio* for an acyloxybenzenesulfonate activator is 1:1, the ratio at which the activator is almost completely converted to the peroxycarboxylic acid [13]. Experiments performed with varying activator-to-bleach ratios have indicated that as the amount of activator is increased, initially the yield of peroxycarboxylic acid also increases. Above an activator-to-peroxygen mole ratio of approximately 1, however, the yield of peroxycarboxylic acid declines. This yield loss is believed to be due to the formation of a diperoxide, which is not a bleaching agent. Diperoxide formation should be minimized.

4.1.5   Peroxycarboxylic Acids

Instead of a bleach activator–peroxygen system in the detergent formulation, which generates the peroxycarboxylic acid as the active bleaching agent in the wash solution, the use of preformed peroxycarboxylic acids has also been studied. While a distinct advantage of these systems is their immediate availability for bleaching when added to the wash solution, the preformed acids have generally been found to be unstable in the alkaline environment typical of detergent formulations. Table 4.1 shows the advantages and disadvantages of the bleach activator–peroxygen systems and the preformed peroxycarboxylic acid systems.

Preformed peroxycarboxylic acids of various types have been developed, and these consist of both monoperoxycarboxylic acids [14–16] (Fig. 4.10) and diperoxycarboxylic acids [17,18] (Fig. 4.11). One of the best known is *diperoxydodecanedioic acid* (DPDDA), which has been found to exhibit superior performance compared with TAED/sodium perborate. DPDDA has been incorporated in a formulated product that has been test marketed in both the United States and Europe.

The performance of preformed peroxycarboxylic acids depends on the pH of the system and the chain length of the peroxycarboxylic acid. As with the activator–peroxygen systems, optimum performance occurs in the pH range 8–9, which is near the pK_a value, while optimum overall cleaning occurs with chain lengths in the range C_6–C_{10} for the monoperoxycarboxylic acids and around C_{12} for the diperoxycarboxylic acids.

Table 4.1   Advantages and Disadvantages of Bleach Activator–Peroxygen Systems Versus Preformed Peroxycarboxylic Acids

Bleach activator–peroxygen systems	
Advantages:	Easier to formulate into laundry product
	Better stability in the formulation
Disadvantage:	In situ reaction required for generation of the peroxycarboxylic acid
Preformed peroxycarboxylic acids	
Advantage:	Can be added directly to the wash solution
Disadvantage:	Unstable in alkaline media

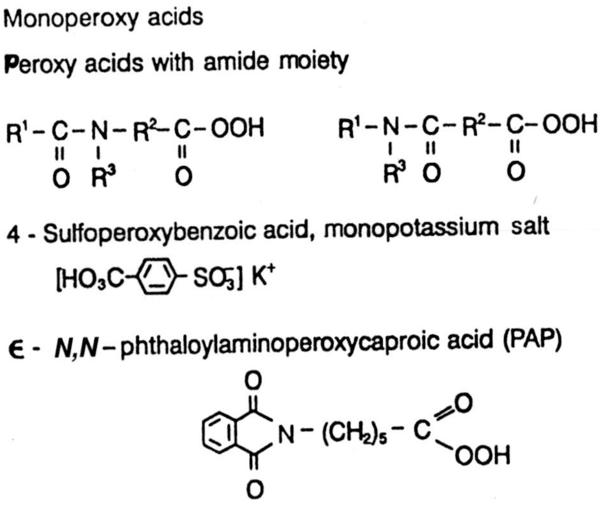

Monoperoxy acids

Peroxy acids with amide moiety

$$R^1-C-N-R^2-C-OOH \qquad R^1-N-C-R^2-C-OOH$$
$$\underset{O \; R^3}{\overset{\parallel \quad |}{}} \qquad\qquad \underset{R^3 \; O}{\overset{| \quad \parallel}{}} \quad \underset{O}{\overset{\parallel}{}}$$

4 - Sulfoperoxybenzoic acid, monopotassium salt

$$[HO_3C-\langle \ \rangle-SO_3^-] \; K^+$$

ε - N,N–phthaloylaminoperoxycaproic acid (PAP)

Figure 4.10   Monoperoxycarboxylic acid.

Diperoxy acids

Substituted diperoxysuccinic acids

$$R-CH-\overset{\overset{\displaystyle O}{\parallel}}{C}-OOH$$
$$\underset{\underset{\displaystyle C}{\overset{\parallel}{}}}{CH_2-O-OOH}$$

Sulfonyldiperoxybenzoic acids

$$HOO-\overset{\overset{\displaystyle O}{\parallel}}{C}-\langle \ \rangle-\overset{\overset{\displaystyle O}{\parallel}}{\underset{\underset{\displaystyle O}{\parallel}}{S}}-\langle \ \rangle-\overset{\overset{\displaystyle O}{\parallel}}{C}-OOH$$

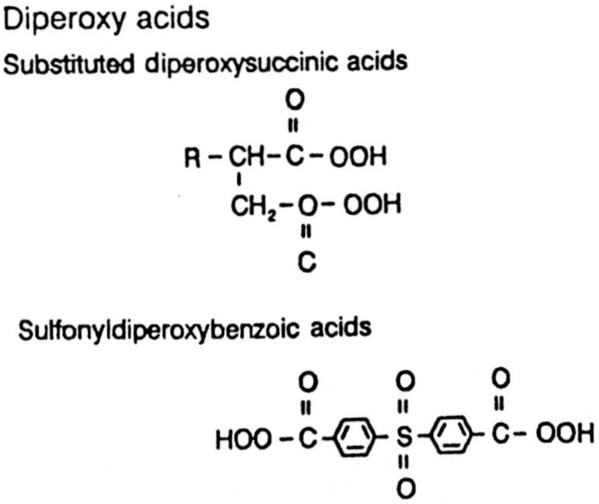

Figure 4.11   Diperoxycarboxylic acids.

As mentioned above, the preformed peroxycarboxylic acids are generally unstable in the alkaline detergent formulations and, therefore, have not been used in detergent formulations. Methods for improving their stability and compatibility with other ingredients in the detergent formulation have been investigated. Some authors have experimented with coatings of various types which, in addition to improving formulation stability, can provide controlled release of the peroxycarboxylic acid into the wash solution [19,20].

Special safety precautions are necessary during the synthesis of preformed peroxycarboxylic acids, which are relatively unstable. Desensitization with inorganic salts such as sodium sulfate is often part of the manufacturing process [21].

4.1.6   Metal-Catalyzed Bleaching Systems

The incorporation of small amounts of certain metal compounds has been found to improve the bleaching performance of peroxygen systems. While these are inexpensive compared to the organic bleach activator–peroxygen systems discussed above, their disadvantages include the possibility of reduced performance in the presence of some detergent builders and water hardness ions. Table 4.2 shows the advantages and disadvantages of these systems.

Various systems containing metal ion compounds have been discussed in the literature [22–29]. As shown in Table 4.3, in metal-catalyzed bleaching systems the use of transition metal ions has been common.

Table 4.2   Advantages and Disadvantages of Metal-Catalyzed Bleaching Systems

Advantages:	Inexpensive system
	Levels of metal activator significantly smaller than needed for organic bleach activator
Disadvantages:	In the presence of detergent builders or water hardness ions, systems are inactivated or have reduced performance
	Possible precipitation of metal oxides and hydroxide under alkaline conditions, which can cause staining of fabrics

Table 4.3   Patents on Metal-Catalyzed Bleaching Systems

Interox

Mixed oxides of calcium and manganese [22]

Manganese compound with calcium, barium, or strontium compound [23]

Cobalt complex metal catalyst [24]

Unilever

Manganese(II) compounds improved by carbonates or metasilicates adsorbed onto solid inorganic support [25,26]

Manganese(II) and copper ions for improving peroxycarboxylic acid bleaching [27]

Transition metal ions in fabric conditioners that catalyze fabric bleaching when subsequently washed with bleach-containing detergent [28]

Procter & Gamble

Mixture of heavy metal cations and an auxiliary metal cation catalyzes peroxygen to different levels [29]

4.1.7   Factors Affecting Bleach Performance, Formulation Stability, and Choice of Bleach System

The factors that affect the performance of a peroxygen bleach system are discussed in Sections 4.1.7.1–4.1.7.8.

4.1.7.1   *Temperature*

Hydrogen peroxide and its perborate and percarbonate derivatives perform best at high temperatures, whereas the peroxycarboxylic acids have optimum performance at the lower temperatures typical of U.S. household washing conditions.

4.1.7.2   *Rate of Solution*

The dissolution rate of the bleach and, in activated systems, of the bleach activator as well, is important for effective bleaching performance. Decreasing particle size and increasing wash temperature increase the dissolution rate of the persalts [30].

4.1.7.3   *Concentration*

Increasing the bleach concentration generally improves stain removal but may affect formulation stability. These factors, together with the cost of the system, need to be evaluated to optimize the bleach concentration in the formulation. Powdered bleach formulations typically contain between 5 and 20% sodium perborate monohydrate, sodium perborate tetrahydrate, or sodium percarbonate, the remainder being sodium carbonate, surfactant, enzyme, optical brightener, and, in some cases, sodium tripolyphosphate. Detergent formulations containing bleach activator usually contain 5–10% sodium perborate monohydrate and 1–2% bleach activator.

4.1.7.4   *pH*

It has been believed that hydrogen peroxide performs most effectively in the high pH range of 10–11 and that the peroxycarboxylic acids have optimum performance in the pH range 8–9, that is, close to their pK_a values [31]. However, studies have indicated that the nature of the stain also determines the optimum pH [4]. For peroxycarboxylic acids, the optimum pH is 7.5–9 for tea stains, while effective removal of grass and red wine stains occurs in the pH range 9–10.

This pH effect also prevents the incorporation of hydrogen peroxide and other peroxygens in aqueous heavy-duty liquid formulations. However, recently, nonaqueous heavy-duty liquid detergent formulations containing bleach and bleach activators have been developed. Further details on such formulations are provided elsewhere [32–34].

4.1.7.5   *Activator/Bleach Ratio*

The activator-to-bleach ratio is important for achieving optimum generation of the peroxycarboxylic acid. For systems in which the stoichiometric ratio for the perhydrolysis reaction is 1, the optimum activator-to-bleach ratio is also close to 1. The presence of excess amounts of the activator can lead to the wasteful generation of diperoxide.

4.1.7.6   Chain Length

Bleaching studies with n-alkyl peroxycarboxylic acids of varying chain lengths have shown that the optimum chain length is the range C_6–C_{10} [4]. While stains such as tea and red wine are effectively bleached by hydrophilic peroxycarboxylic acids, grass stains are more effectively removed by longer chain hydrophobic peroxycarboxylic acids. Therefore, for maximum performance, a mixture of peroxycarboxylic acids with a range of chain lengths may be preferable to a single peroxycarboxylic acid having a fixed chain length.

4.1.7.7   Stains

Stains remaining in the fibers generally are more difficult to bleach than stains in solution. Different stains also have different affinities for fibers of various types, which, in turn, affect the bleach performance. For example, tea stain does not have an affinity for polyester fibers and is desorbed rapidly from the fibers into the wash solution. However, tea stain has a strong affinity for nylon and is difficult to bleach from that fabric.

4.1.8   Formulating Bleaches

The formulation of bleach systems in a detergent requires the consideration of such factors as:

*   Moisture in detergent formulation
*   Nonadsorbed surfactant
*   Moisture permeability of detergent container
*   Temperature of the detergent base
*   Coating and stability of the bleach and activator
*   Particle size of detergent ingredients

The presence of moisture reduces bleach stability, resulting in the loss of available oxygen of the system. Moisture can be present due to other ingredients in the detergent base or due to permeability of the detergent container. The presence of a bleach activator also can affect the stability of the peroxygen bleach. Methods such as coating the activator are used to reduce this loss.

Among the factors that affect the selection of the bleach system for a detergent formulation are:

*   Effectiveness of bleach system on bleachable as well as on other soils present in laundry
*   Cost
*   Availability of the bleach system
*   Compatibility of the bleach with other detergent ingredients

Although the primary function of the bleach is to remove bleachable soils, overall cleaning must be provided as well. The cost and availability of the bleaching system are important considerations. Activator–peroxygen bleach systems are currently used in commercially available detergents, since they are stable in the formulations. In the future, improved methods of stabilizing or coating preformed peroxycarboxylic acids may allow incorporation of these agents into detergent formulations.

4.1.9   Performance

The performance of a system is usually evaluated by washing stained swatches in a washing machine or a Terg-o-tometer. The stain removal is evaluated by measuring the reflectance of the swatch before and after washing.

Table 4.4 compares the stain removal effectiveness for detergent containing no bleach, detergent with added formulated bleach, detergent containing activator–peroxygen bleach, detergent with added bleach containing the peroxycarboxylic acid DPDDA, and detergent with added hypochlorite bleach. The results indicate that bleach contributes to overall cleaning because it removes not only the bleachable soils but also detergeable soils. The detergent containing activator–bleach is equivalent in performance to the detergent with DPDDA added, and both are superior to that of detergent alone, or detergent with only peroxygen bleach added. Hypochlorite bleach is most effective in terms of bleaching performance but causes the greatest fiber damage.

Table 4.5 compares two activated bleaching systems, one containing TAED and the other containing acyloxybenzenesulfonate as the activator. Both systems contained sodium perborate monohydrate. The surface active properties of the latter system contribute to its superior performance.

Table 4.4   Stain Removal Performance of Detergent–Bleach Systems

Stain	Stain removal (%) in five systems[1]				
	A	B	C	D	E
Red wine	72	70	79	87	99
Coffee	80	86	84	88	102 [2]
Tea	28	53	63	83	103
Grass	82	93	95	90	112
Oil/charcoal	6	12	11	11	9
Spangler sebum	63	75	73	65	76
Makeup	10	16	13	13	13
Mustard	75	94	97	97	102
Ketchup	77	89	98	94	102

[1]A = 1.5 g/L detergent I (contains phosphate and enzyme); B = 1.5 g/L detergent I + 2 g/L peroxygen bleach containing enzyme; C = 1.5 g/L detergent containing acyloxybenzene bleach activator, sodium perborate monohydrate, phosphate, enzyme (available oxygen: 2.8 ppm from peroxy-carboxylic acid and 20.9 ppm total); D = 1.5 g/L detergent I + 1.2 g/L bleach containing 8.9% diperoxydodecanedioic acid (13.6 ppm available oxygen); E = 1.5 g/L detergent I + 5 g/L hypochlorite bleach.
[2]Stain removal performance is greater than 100% in some cases due to the attainment of whiteness greater than that of the original fabric.
Source: Ref. [3].

Table 4.5   Stain Removal Performance of Bleach Activators

Stain	Stain removal (%) in three systems[1]		
	A	B	C
Red wine	37	40	43
Tea	24	34	45
Motor oil	32	25	35
Gravy	66	68	78

[1]A = 2.16 g/L detergent II (contains phosphate, enzyme, and sodium perbo-rate monohydrate); B = 2.16 g/L detergent II + 0.09 g/L TAED bleach activator; C = 2.16 g/L detergent II + 0.09 g/L acyloxy-benzenesulfonate bleach activator.
Source: Ref. [3].

4.2   Optical Brighteners

4.2.1   Introduction

Optical brighteners or *fluorescent whitening agents* are used to provide optical compensation to enhance the brightness and whiteness of laundry. These organic compounds absorb the invisible ultraviolet component of light and emit this energy in the longer wavelength blue spectrum of visible light. White fabrics that have a blue tinge as a result of the addition of optical brighteners are perceived as having improved whiteness and brightness.

Optical brighteners must be optically colorless on the fabric surface; they also must not absorb light in the visible part of the spectrum [35]. Optical brighteners are dyes that have some reflectance in the visible portion of the spectrum in addition to their emission spectrum. With repeated laundering, accumulation of optical brighteners on the fabric surface can result in emission in the visible range of light, perceived as discoloration of the fabric [1].

Optical brighteners are introduced into the washing solution as an ingredient of the laundry detergent formulation. For the brightener to be effective, it must dissolve or disperse in the washing solution and then become distributed on the fiber surface and throughout the fiber. The effectiveness of brighteners is therefore determined by their rates of dissolution in the washing solution and their distribution on the fibers. These are determined by the chemical nature of the optical brightener, the washing conditions, and the types of fiber present [36].

4.2.2   Types of Optical Brighteners

The four main types of optical brighteners are [1,35]:

* Optical brighteners with affinity to cotton
* Optical brighteners with affinity to polyamide
* Optical brighteners with affinity to polyester
* Hypochlorite-resistant optical brighteners

Various theories have been proposed to explain the affinity of optical brighteners for different types of fiber [37]. Cotton and hypochlorite-resistant optical brighteners are believed to attach themselves to the fabric surface by means of hydrogen bonding. The effectiveness of polyamide and polyester optical brighteners appears to depend on their diffusion into the fiber surface. The chemical structures of some commercially available optical brighteners and their affinities to different types of fibers are shown in Figs. 4.12–4.15.

The most commonly used *cotton optical brighteners* are bistriazinyl derivatives of 4,4′-diaminostilbene 2,2′-disulfonic acid. Figure 4.12 shows the four most commonly used anionic cotton brighteners, listed from left to right in order of increasing water solubility [36].

The general structures of typical *polyamide optical brighteners* are shown in Fig. 4.13. These are amino coumarin and diphenyl pyrazoline derivatives, which are not stable in the presence of hypochlorite bleach. These optical brighteners are typically used in light-duty detergent formulations [37].

Figure 4.14 illustrates structures of naphthotriazolyl and bisbenzoxazolyl derivatives which have good affinity to polyester and nylon besides having good stability in the presence of hypochlorite bleach [37].

The chemical structure of a commonly used *hypochlorite-bleach-stable optical brightener* is shown in Fig. 4.15. This is a derivative of naphthotriazolylstilbenesulfonic acid.

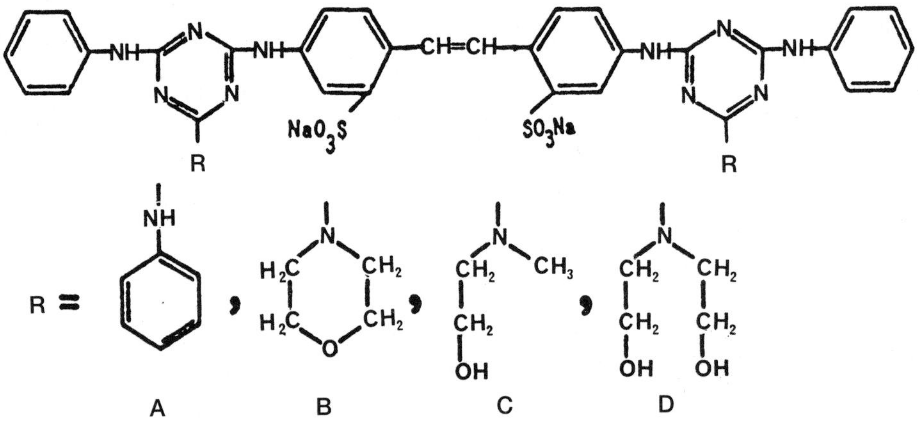

Figure 4.12 The four most commonly used anionic cotton brighteners: water solubility increases from A to D.

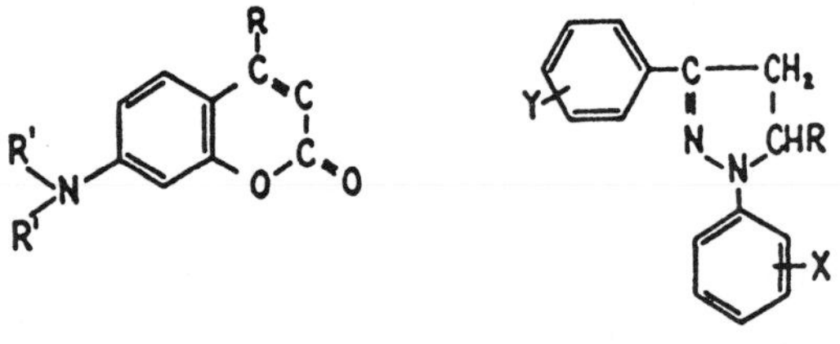

Figure 4.13 Polyamide optical brighteners derived from amino coumarin (left) and diphenyl pyrazoline (right).

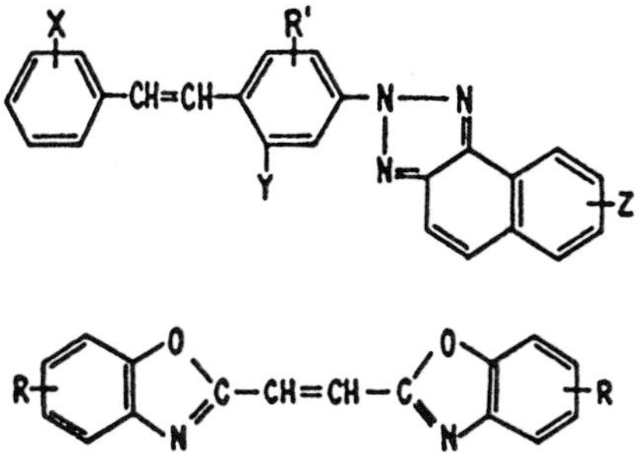

Figure 4.14   The naphthotriazolyl (top) and bisbenzoxazolyl (bottom) molecules.

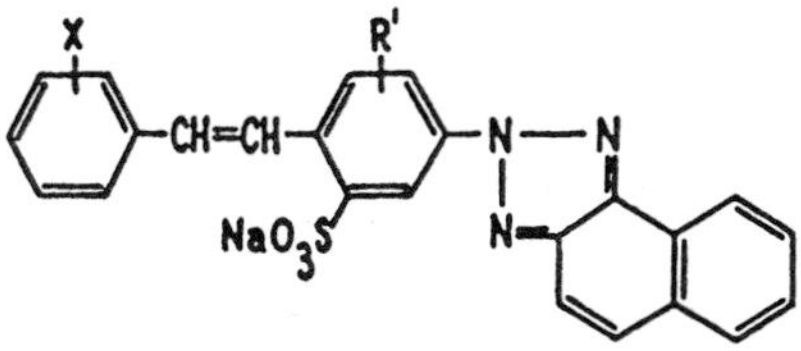

Figure 4.15   Optical brightener that is stable in hypochlorite bleach: derived from naphtho-triazolylstilbenesulfonic acid.

While all optical brighteners were considered to be stable with oxygen bleach, that is, in the presence of sodium perborate, the use of bleach activators and peroxycarboxylic acids in laundry products has limited the selection of optical brighteners that can be used. The structure shown in Fig. 4.15 can be used with these systems, but it may be slowly destroyed by certain formulations [38]. Research is under way toward the development of new compounds with improved stability. Recently, a benzofuranyl biphenyl compound was developed as the optical brightener for detergent products containing a peroxycarboxylic acid or a bleach–activator system [39].

4.2.3   Performance and Selection of Optical Brighteners

The various factors that determine the performance and selection of optical brighteners [36–38,40–42] are discussed in Sections 4.2.3.1–4.2.3.15.

4.2.3.1   Laundry Product Composition [35]

The performance and selection of the optical brightener depends on the type and amount of surfactant, builder, and other components in the laundry product.

The type and structure of the surfactant present in the detergent determines the effectiveness of the optical brightener. Anionic surfactants are generally found to be compatible with anionic optical brighteners.

In the presence of nonionic surfactants, which are commonly found in liquid detergent formulations, optical brightener performance is reduced to differing degrees depending on the degree of ethoxylation of the surfactant. These effects are believed to be due to the formation of mixed micelles comprising surfactant and optical brightener [40]. However, the addition of anionic surfactant is found to improve the performance.

Cationic surfactants in fabric softeners neutralize the anionic charge of the optical brighteners, leading to the loss of the important optical brightener properties of absorption of UV light and emission of visible light.

Now we consider the effect of surfactant type on the performance of four cotton optical brighteners (A, B, C, D, in order of increasing water solubility), whose chemical structures are shown in Fig. 4.13. The experimental conditions were as follows.

The washing tests were performed with cotton swatches for 10 minutes at 130 °F in an unbuilt system using 0.5 g/L of active linear alkylbenzenesulfonate anionic surfactant, ethoxylated nonylphenol nonionic surfactant, or dihydrogenated tallowdimethylammonium chloride cationic surfactant. The pH of the wash solution was in the 5.5–7 range. The concentration of each optical brightener used was selected to provide the same fabric fluorescence as in the anionic system. Similar tests were done for a built system with the addition of 1 g/L of sodium tripolyphosphate (STPP: pH 9.5). After washing, the fabric swatches were air-dried in the dark and the fluorescence values of the swatches were measured. Fabric fluorescence is generally a physical measurement of the amount of optical brightener exhausted onto the fabric [36].

The data in Fig. 4.16 show how optical brightener performance is reduced in the presence of nonionic and cationic surfactants. Optical brightener A is thus recommended for systems containing anionic but not nonionic surfactant, while optical brighteners B, C, and D can be used for both anionic and nonionic systems. Optical brightener D can also be used with cationic fabric softeners [37].

As shown in Fig. 4.17, a built system containing STPP exhibits improved optical brightener effectiveness with nonionic surfactant relative to the unbuilt system shown in Fig. 4.16. The electrolyte may decrease brightener dissociation and increase the concentration of ion pairs, leading to improved optical brightener deposition on the fabric.

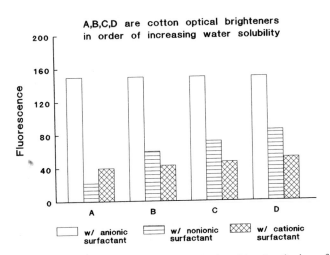

Figure 4.16   Optical performance in the presence of nonionic acid and cationic surfactants: unbuilt systems.

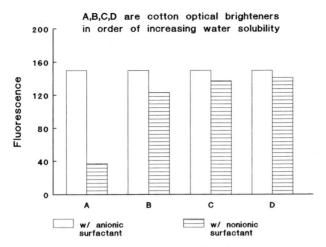

Figure 4.17   Optical performance in the presence of different surfactants: systems built with sodium tripolyphosphate.

The interaction of some optical brighteners with strongly alkaline builders has been found to cause coloration in some laundry products [43]. The formulator can enhance the color of the product, or prevent coloration, by selecting the appropriate optical brightener and alkaline builder, and by carefully choosing their order of addition during the blending process.

For laundry products that contain bleach activators, peroxycarboxylic acids, or hypochlorite bleaches, the selection of optical brighteners is limited. Compounds having the structure shown in Fig. 4.15 can be recommended for such products, but these may be slowly destroyed when incorporated in certain formulations.

4.2.3.2   Laundry Product Type

Optical brighteners are used in powder and liquid heavy-duty detergents, light-duty detergents for fine fabric, dry oxygen and chlorine bleaches, and fabric softeners. Heavy-duty detergents for mixed fabric loads and for fine fabric detergents require optical brighteners of different types [38].

For liquid detergents, optical whitener selection is determined by solubility properties as well as compatibility with nonionic surfactants. Another factor to be considered with liquid detergents involves their use as prespotters or stain removers, applied before general washing of the fabric. The application of a highly concentrated optical brightener solution to a limited area on the fabric can leave behind a fluorescent stain. Studies have indicated that optical brighteners having the structure show in Fig. 4.16 and the types A and B in Fig. 4.13 minimize this staining effect.

With cationic surfactants present in fabric softeners, the most soluble anionic optical brighteners are used. They have reduced effectiveness in these formulations because their inherent incompatibility with the cationic surfactant leads to the quenching of fluorescence by the cationic surfactant. The lower performance of optical brighteners in rinse-additive fabric softeners is usually acceptable, since the primary fabric whitening is accomplished in the wash cycle.

For combination detergent/softener products, the optical brightener must perform effectively in the presence of cationic surfactant. A compound of the type D (Fig. 4.13) can be used for such powdered formulations. In the case of liquid detergent/softener products that contain nonionic and cationic surfactants, the currently used optical brighteners are ineffective. An amphoteric distyrylbenzene optical brightener that performs effectively in such formulations has recently been developed [44].

4.2.3.3   Ease of Incorporation into the Formulation

The physical form of the optical brightener should facilitate the incorporation of the additive into the product. The bead or granular forms typically used reduce exposure to airborne dusts and also, because of their flowability, allow the use of the automated solids metering devices common in detergent plants. Proper particle size is important for satisfactory mixing and to prevent separation in dry-blended formulations. Good solubility or dispersibility is required for liquid formulations.

4.2.3.4   Solubility of the Optical Brightener [41]

The commonly used optical brighteners generally exhibit low total solubility, which enhances their deposition on the fabric surface. However, for liquid formulations, this low solubility may limit optical brightener usage unless stable dispersions of the optical brightener can be incorporated.

4.2.3.5   Rate of Solution of the Optical Brightener [41]

Rate of solution is important primarily for powdered laundry products that are used at low washing temperatures. While the optical brightener may be completely soluble under the washing conditions, the time required for complete solubilization at low temperatures may be greater than the typical washing time of 10 minutes. Thus, the total whitening effect of the optical brightener may not be achieved.

4.2.3.6   Laundry Product Whitening [42]

An optical brightener is required primarily to enhance the appearance of white fabrics in the laundry. A secondary function is to improve the appearance of the laundry product itself. Some types of optical brightener, and blending techniques used for their incorporation in the powder formulation, may result in product discoloration. Therefore, the selection of the optical whitener, as well as the order of addition during detergent formulation, are important for achieving optimum powder whiteness and performance.

4.2.3.7   Washing Conditions

The ratio of fabric to wash solution, the degree of agitation, and the temperature affect the time required for the optical brightener to dissolve completely in the wash solution. Optical brighteners that dissolve quickly at low temperatures will be selected for cold water detergent formulations.

4.2.3.8   Fabric Types Present in Wash Load and Optical Brightener Affinity to Fabric Surface

Optical brighteners are compounds that function by adsorbing on the fabric surface. Therefore, one of the most important factors that affects performance is substantivity to the fabric surface. The fiber composition of the laundry needs to be considered, as well as the interaction between the brighteners used on the fabric at the textile mill and the optical brighteners in the laundry detergent.

4.2.3.9   Effect of Soil

The presence of soil on the fabric reduces optical brightener effectiveness. Soil redeposited on the fabric absorbs light and reduces optical brightener performance. Therefore, for the optical brightener to be effective, the detergent must be able to first remove the soil.

4.2.3.10   Bleach Stability

Optical brighteners that are stable in the presence of hypochlorite bleach and activated peroxygen bleach systems have been discussed. The bleach stability of optical brighteners needs to be evaluated under product storage conditions as well as under typical washing conditions.

4.2.3.11   Fastness of the Optical Brightener

Optical brighteners on fabrics may be exposed to light and air, to heat during the washing and drying cycles and during ironing of the fabric, and to perspiration, deodorants, and other substances. Therefore, the fastness of the optical brightener should be evaluated under such exposures.

4.2.3.12   Use Concentration of the Optical Brightener

The amount of optical brightener to be used in the detergent product affects its performance and selection. A particular optical brightener may have better properties and exhibit improved whiteness at low concentrations than at higher levels. Excess buildup of optical brightener on the fabric results in an undesirable yellow appearance. Detergent formulations typically contain optical brightener levels between 0.05 and 0.5%.

4.2.3.13   Toxicity

The toxicity of the optical brightener selected needs to be evaluated, in particular, for skin irritation and photoallergenic properties, and effect on the environment. Also to be determined is the effect of other ingredients in the detergent formulation on the toxicity of the optical brightener.

4.2.3.14   Hue of Whiteness Produced

The type or hue of whiteness the selected optical brightener will produce on the fabric is another important factor. Both visual and instrumental evaluations are performed [37].

4.2.3.15   Cost

The cost of the optical brightener is another important factor in the selection of the compound. Optical brighteners that provide the best overall performance at the lowest cost will be chosen.

References

1. Coons, D., Dankowksi, M., Diehl, M., Jakobi, G., Kuzel, P., Sung, E., Trabitzsch, U., in *Surfactants in Consumer Products—Theory, Technology, and Application*, J. Falbe, Ed., Springer-Verlag, Heidelberg, 1987.
2. *A Bleacher's Handbook*, Solvay Interox publication, Houston, TX.
3. Parker, J.A., Raney, O.G., The use of peroxygen bleaches as a means of improving detergent performance, presented at the New Horizons Conference, Hershey, PA, October 1989, Solvay Interox publication, Houston, TX.
4. James, A., Chemistry of peroxygen bleaching, presented at the 83rd Annual Meeting of the American Oil Chemists' Society, Toronto, Ont., Canada, May 1992, Solvay Interox publication, Houston, TX.
5. Kao, Japanese Patent 59,193,999.
6. Procter & Gamble, U.S. Patent 4,412,934.
7. Akzo, European Patent Application 0210674.
8. Clorox, European Patent Application, 0267048.
9. Clorox, European Patent Application, 0252724.
10. Clorox, European Patent Application 0262895.
11. Lever, U.S. Patent 4,397,757.
12. Kao, European Patent Application 0284292.
13. Kissa, E., Dohner, J.M., Gibson, W.R., Strickman, D., Kinetics of staining and bleaching, J. Am. Oil Chem. Soc., *68* 532 (July 1992).
14. Procter & Gamble, U.S. Patent 4,634,551.
15. Interox, European Patent Application 0212913.
16. Hoechst, European Patent 349940.
17. Monsanto, European Patent Application 083560.
18. Monsanto, European Patent Application 026175.
19. Procter & Gamble, U.S. Patent 4,391,723.
20. Procter & Gamble, U.S. Patent 4,374,035.
21. Degussa, Henkel, European Patent Application 127783 A.
22. Interox, European Patent Application 0196738.
23. Interox, U.S. Patent 4,620,935.

24. Interox, European Patent Application 0272030.
25. Unilever, European Patent Application 082563.
26. Unilever, U.S. Patent 4,538,183.
27. Unilever, European Patent Application 0143491.
28. Unilever, European Patent Application 0257860.
29. Procter and Gamble, European Patent Application 072166.
30. Shehad, N.S., Espinosa, J.T., Bui, K., Factors affecting dissolution rate of persalts, Solvay Interox publication, Houston, TX.
31. James, A.J., MacKirdy, I.S., The chemistry of peroxygen bleaching, Chem. Ind., *15* 641 (Oct. 15, 1990).
32. Hepworth, P., Heavy duty laundry liquids with bleach—The way for the nineties, HAPPI, 76 (May 1992).
33. INFORM, *3*(3) 293 (1992).
34. Broze, G., Concentrated liquid laundry detergents, HAPPI, 62 (June 1992).
35. Zweidler, R., Hefti, H., Fluorescent brighteners, in *Kirk–Othmer Encyclopedia of Chemical Technology*, Vol. 4, 3rd ed., Wiley, New York, 1978.
36. Stensby, P.S., The effect on brightener performance of various surfactants, Deterg. Age (Jan.–Feb. 1968).
37. Stensby, P.S., Optical brighteners and their evaluation, Soap Chem. Spec. (April–Sept. 1967).
38. Findley, W.R., Fluorescent whitening agents for modern detergents, J. Am. Oil Chem. Soc., *65* 679 (April 1988).
39. U.S. Patent 5,089,166.
40. Schussler, U., Sewekow, U., The effect of nonionic surfactants on the affinity of fluorescent whitening agents for detergents, presented at the SEPAWA Conference, Bad Durkheim, Germany, 1977.
41. Findley, W.R., Whitener selection for today's detergents, J. Am. Oil Chem. Soc., *60* 1367 (July 1983).
42. Neiditch, O., Fluorescent whitening agents—Still an important technology, HAPPI, 86 (Nov. 1992).
43. Lange, K.R., Fluorescent whitener interactions with alkaline builders, Deterg. Spec., *6* 19 (1969).
44. U.S. Patent 4,478,598.

CHAPTER 5

Polymers in Cleaners

David Witiak

Polymers have been used in cleaners for the past 15–20 years, and their presence continues to grow. This chapter reviews the types of polymer used in cleaners, the benefits they contribute, and their application in the formulation of products. Use of polymers in household laundry, industrial and institutional (I&I) laundry, industrial dishwashing, and hard-surface cleaners, with emphasis on the U.S. market, is covered. Polymers are widely used in cleaner formulations worldwide.

5.1   Polymer Types

In the detergent area two carboxylate polymers dominate overall usage:

Acrylic acid homopolymer (pAA)
Acrylic acid/maleic acid copolymer (pAA/MA)

$$-(CH-CH_2)_x- \qquad\qquad -(CH-CH_2)_x-(CH-CH)_y- \\ \quad | \qquad\qquad\qquad\qquad\qquad | \qquad\quad | \quad\; | \\ \quad COOH \qquad\qquad\qquad COOH \quad HOOC \;\; COOH \qquad (5.1) \\ \\ \qquad pAA \qquad\qquad\qquad\qquad\qquad pAA/MA$$

These polymers are used primarily to improve overall cleaning performance. For some formulations or manufacturing processes (e.g., spray-drying or agglomeration), polymers can also improve detergent formulation processing. The largest use of polymers occurs in home laundry: indeed, almost all major powdered laundry detergent formulations contain a polymer. The predominant polymers used in home laundry in the United States are acrylic acid homopolymers, while in Europe, pAA/MA copolymers are predominantly used. In the I&I market, acrylic acid homopolymers are used the most. Other polymer types find use in selected formulations: for example, sodium carboxymethylcellulose (CMC), poly(vinyl-pyrrolidone), poly(methyl vinyl ether) maleic acid, hydrophobe/maleic acid copolymers, and hydrophobe/acrylic acid copolymers. CMC has been used for many years as an antiredeposition agent for particulate soil on cotton. With the incorporation of carboxylate polymers into detergent formulations, the use of CMC is declining.

David Witiak, Rohm & Haas Corporation, Spring House, Pennsylvania 19477, U.S.A.

5.2   Polymer Functions

Carboxylate water-soluble polymers perform various functions in a detergent formulation. The key ones are discussed in Sections 5.2.1–5.2.4.

5.2.1   Preventing Inorganic Salt Precipitation

In the late 1940s detergent manufacturers began using sodium tripolyphosphate (STPP) as the main builder in powder laundry detergents. In the early 1970s legislation limiting the use of phosphates started the influx of new "no-phosphate" detergents. Several of these new detergents contained soda ash as the main builder. Soda ash removes hardness ions (mainly calcium) via the formation of insoluble calcium carbonate. However, in areas having high hardness levels, the formation of calcium carbonate leads to graying and stiffening of fabrics. The buildup of calcium carbonate on clothes is called encrustation. In machine dishwashing, scale that forms on spray nozzles and machine surfaces can show up as filming and spotting on tableware.

Figure 5.1   Calcium carbonate crystals without additive (top) and modified with pAA (MW 4500) (bottom). Magnification, 5000×.

Carboxylic acid based polymers affect the deposition of calcium carbonate by acting as crystal growth modifiers and dispersants for the growing crystal nuclei. These polymers adsorb onto the growing crystal face, causing distortion (Fig. 5.1), and preventing further growth. The distorted particles have less tendency to adhere to fabric or hard surfaces.

In fabric encrustation, the calcium carbonate crystals preferentially deposit on cotton fibers. The effect of polymer molecular weight on encrustation of a 40–50% soda ash built home laundry detergent is show in Figs. 5.2 and 5.3. Encrustation is minimized when

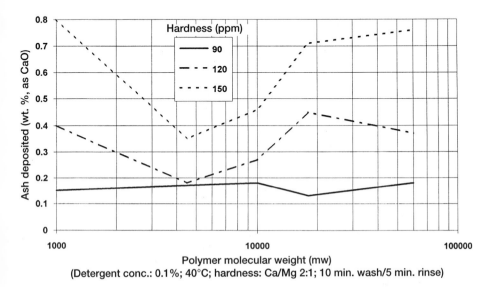

(Detergent conc.: 0.1%; 40°C; hardness: Ca/Mg 2:1; 10 min. wash/5 min. rinse)

Figure 5.2   Effect of polymer molecular weight on encrustation of cotton fabric.

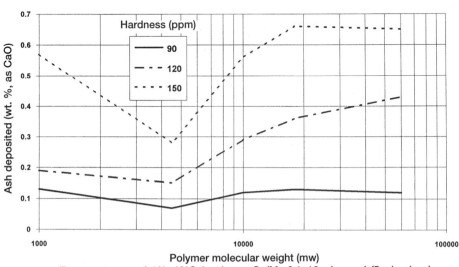

(Detergent conc.: 0.1%; 40°C; hardness: Ca/Mg 2:1; 10 min. wash/5 min. rinse)

Figure 5.3   Effect of polymer molecular weight on encrustation of polyester/cotton (65:35) fabric.

polymer molecular weight is about 4500. As water hardness increases, the tendency for calcium carbonate to deposit increases. During the washing process, consumers can minimize encrustation by filling the washing machine, adding the detergent, and allowing the detergent to dissolve and react with the hardness ions before adding clothes. If detergent is added to the machine with water and laundry already there, the calcium carbonate will nucleate on the fabric surface and encrustation will result.

5.2.2   Increasing Soil Removal

During laundering, surfactants loosen oily and particulate soils from the fabric surface. Polymers aid this process by adsorbing onto the loosened soil, dispersing it into the wash bath. The effect of polymer molecular weight on clay soil removal and redeposition is shown in Figs. 5.4 and 5.5, respectively. As polymer molecular weight increases, flocculation of soil onto fabric surface occurs, hindering soil removal. Proper selection of polymers is key to enhancing performance.

5.2.3   Preventing Particulate Soil Redeposition

During laundering or dishwashing, soil is removed from the fabrics or hard surfaces and concentrates in the wash bath. Polymers adsorb onto the particulate soil surface, increasing its overall negative charge and reducing the tendency of the soil to redeposit on the cleaned surfaces. The effect of polymer molecular weight on soil redeposition is shown in Fig. 5.6. For preventing particulate soil redeposition, acrylic acid homopolymers (MW 4000–10,000) give the best performance.

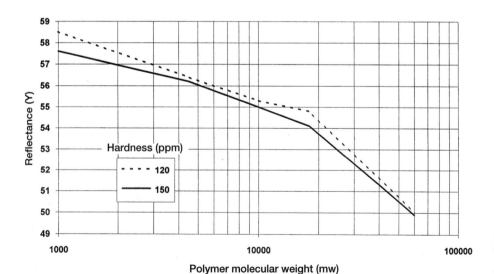

Figure 5.4   Effect of polymer molecular weight on the removal of clay soil from cotton.

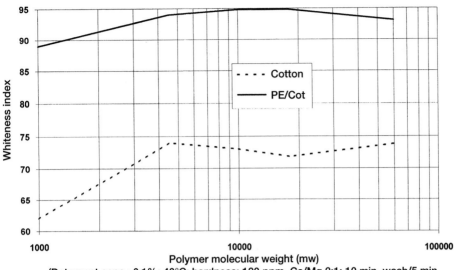

Figure 5.5   Effect of polymer molecular weight on the redeposition of clay on cotton and polyester/cotton fabrics.

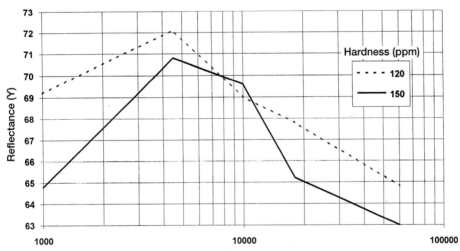

Figure 5.6   Effect of polymer molecular weight on the removal of clay soil from polyester/cotton (65:35) fabrics.

5.2.4   Other Benefits

5.2.4.1   Soil Release

Oily soils are difficult to remove from polyester-containing fabrics because these fibers are oleophilic and oily stains adhere strongly to their surfaces. In addition, polyester is hydro-

Table 5.1   Effect of Soil Release Polymer on Cleaning Performance

Soil release polymer (wt %)[1]	Soil release (%)[2]
0	29.4
2	87.4

[1]Nonionic graft copolymer. (*Note*: Polymer is subject to hydrolysis in highly alkaline systems.)
[2]Temperature, 35 °C; hardness, 150 ppm; soil, detergent motor oil; fabric, polyester doubleknit; 16% linear alkylbenzene-sulfonate; 7% nonionic surfactant, 2% triethanolamine, 6% sodium xylene sulfonate.

phobic, making it difficult for surfactants to wet the surface. Modifying the fabric surface to make it more hydrophilic should enhance cleaning performance. The enhanced cleaning performance obtained when polyester fabric is washed with a liquid laundry detergent containing a nonionic graft copolymer (soil release polymer) is shown in Table 5.1 [1]. In these tests the fabric is prewashed with the detergent, then soiled, and washed using the same liquid laundry formulation used in the prewash step.

5.2.4.2   Dye Transfer Inhibition

During washing, dyes released from dark fabrics can interact with lighter or white fabrics, imparting a distinct color or tint. Polymers such as poly(vinylpyrrolidone) or poly(vinyl-imidazole) can form strong complexes with a wide variety of dyes in aqueous solution, to prevent them from transferring to fabrics during washing. The complexes are usually formed through hydrogen bonding. However, in certain cases other forces, such as hydrophobic attraction, van der Waals forces, and charge–transfer interactions, may also occur [2].

5.3   End-Use Applications

5.3.1   Home Laundry

In laundry applications, removing soil and preventing soil redeposition are of primary concern. U.S. consumers use about 5 billion pounds of home laundry detergents per year. Most powdered laundry detergents currently available contain polymers [e.g., CMC, acrylic acid homopolymers, acrylic acid/maleic acid polymers, poly(ethylene glycol) (PEG)]. CMC has been used for many years to prevent particulate soil from depositing on cotton. Modified cellulosics act as soil release agents, particularly on fabrics containing polyester [3]. Combinations of low molecular weight acrylic acid homopolymers and poly(ethylene glycol) substantially increase particulate soil removal in zeolite/soda ash based detergents (Table 5.2) [4].

Polymers are added to soda ash based detergents where encrustation is a more serious concern; they not only help reduce encrustation, they can also aid in the removal of particulate soil and in the prevention of soil redeposition (Table 5.3).

Table 5.2   Effect of Polymer on Clay Soil Removal and Antiredeposition Benefits in a Zeolite/Soda Ash Built Detergent[1] Used on Cotton and Polyester/Cotton (PE/Cot) Fabrics

| Polymer system (MW) | | Whiteness index | | |
| | | Clay removal[2] | | Antiredeposition[3] |
pAA (4500)	PEG (8000)	Δ Cotton	ΔPE/Cot	PE/Cot
2.4	0	-4.1	-2.0	103
1.8	0.6	0	-1.1	116
1.2	1.2	-0.7	0	116
0.6	1.8	-0.9	-1.1	115
0	2.4	-5.0	-4.1	110

[1]Temperature, 35 °C; hardness, 120 ppm; 7.5% linear alkylbenzenesulfonate, 7.5% alkyl sulfate, 2% nonionic surfactant, 24% zeolite, 13% soda ash, 20% sodium sulfate, 1% sodium silicate.
[2]0 = best performance; Δ = difference between best and other samples.
[3]Higher values = better performance.

Table 5.3   Effect of Polymer on Clay Soil Removal, Antiredeposition, and Encrustation in a Soda Ash Based Detergent[1] Used on Cotton (Cot) and Polyester/Cotton (PE/Cot) Fabrics

| Polymer | Encrustation reduction | | Clay removal (reflectance) | | Clay redeposition (whiteness index) | |
	Cot	PE/Cot	Cot	PE/Cot	Cot	PE/Cot
None			54.7	69.8	46	74
1% pAA[2]	60%	50%	56.0	72.1	74	94

[1]Temperature, 40 °C; hardness, 120 ppm; 0.1% detergent, 40–50% soda ash/nonionic surfactant/sodium sulfate, < 1% CMC.
[2]MW 4500.

Polymers can also improve clay soil removal and help reduce soil redeposition in formulations containing tetrasodium pyrophosphate (TSPP) or low levels of STTP (Table 5.4). Results show that the higher molecular weight acrylic acid homopolymer reduces soil redeposition but decreases soil removal compared with the lower molecular weight acrylic acid homopolymer or the high molecular weight acrylic acid/maleic acid copolymer. If the formulation contains TSPP, polymer can also reduce the encrustation formed by this builder in hard water.

5.3.2   Dishwashing

5.3.2.1   I&I Dishwashing

In industrial and institutional (I&I) dishwashing, highly alkaline liquid formulations containing tetrapotassium pyrophosphate (TKPP), potassium hydroxide (KOH), and potassium

Table 5.4   Effect of Polymer on Clay Soil Removal and Antiredeposition in a Low Phosphate Detergent[1] Used on Cotton and Polyester/Cotton (PE/Cot) Fabrics

Polymer (MW)	Molecular weight	Clay removal (reflectance)		Clay redeposition (whiteness index)	
		Cotton	PE/Cot	Cotton	PE/Cot
None		58.6	65.1	70	92
pAA	4,500	59.7	65.9	86	98
pAA	60,000	56.5	61.8	83	98
pAA/MAA[2]	3,500	62.4	66.2	84	99
pAA/Maleic acid	70,000	60.0	65.2	83	99

[1]Temperature, 25 °C; hardness, 200 ppm; 0.2% detergent.
[2]Acrylic acid/methacrylic acid copolymer.

Table 5.5   Relative Cost Performance of Phosphate/Polycarboxylate Based Formulations[1]

TKPP	Polymer[2]	Cost[3]	Filming[4]	Spotting[4]
20%	None	20	2	0.5
10%	None	10	2	2.5
10%	2%	12	1	0.5
10%	4%	14	1	0.5
10%	6%	16	0.5	0.5

[1]Temperatures: wash, 71 °C; rinse, 88 °C; hardness, 200 ppm; soil, 60 g (margarine/dry milk, 80:20); 10% KOH, 2% potassium silicate, 2% NaOCl; 20 cycles.
[2]Polyacrylic acid (MW 4500).
[3]Relative: TKPP, $1/solid lb; polymer $1/solid lb [5].
[4]0 = no filming or spotting; 4 = entire surface covered with film or spots.

silicates are commonly used. These formulations provide good cleaning and result in spotless, film-free glasses and flatware. The use of potassium salts yields clear and stable formulations. Polycarboxylate polymers have been used in these formulations since the early 1970s. With the use of polymers, builder levels can usually be lowered, thereby reducing raw material cost, while at the same time maintaining overall performance (Table 5.5) [2–5].

The filming and spotting benefits of acrylic acid polymers can also be observed in no-phosphate formulations based on sodium nitriloacetate (NTA)/silicate/caustic and silicate/caustic (Tables 5.6 and 5.7, respectively) [6].

5.3.2.2   Home Dishwashing

Current home dishwashing formulations are based on STPP, soda ash, and silicates. The high level of STPP (typically > 30%) in these formulations provides excellent dispersion and sufficient hard water control to minimize filming and spotting. In the early 1990s, nonphosphate (and nonchlorinated) home dishwashing detergents were introduced into the European market, primarily in response to environmental pressures. These new detergents use sodium

Table 5.6   Performance of Sodium Nitrilotriacetate (NTA)/Polycarboxylate Based Formulations[1]

NTA	Polymer[2]	Filming[3]	Spotting[3]
30%	None	0.5	1.5
12%	None	1.5	2
9%	None	2.5	2.5
9%	4%	1.5	0.5
12%	4%	1	0.5

[1]Temperatures: wash, 71 °C; ;rinse, 88 °C; hardness, 200 ppm; soil, 60 g (margarine/dry milk, 80:20); 8% NaOH, 14% sodium silicate (N, from PQ Corp.); 20 cycles.
[2]Polyacrylic acid (MW 4500).
[3]0 = no filming or spotting; 4 = entire surface covered with film or spots.

Table 5.7   Performance of Silicate/Caustic/Polycarboxylate Based Formulations[1]

Polymer[2]	Filming[3]	Spotting[3]
None	4	4
2%	2	1.5
4%	2	1
6%	1.5	1

[1]Temperatures: wash, 71 °C; rinse, 88 °C; hardness, 200 ppm; soil, 60 g (margarine/dry milk, 80:20); 10% NaOH, 6% sodium silicate (BW, from PQ Corp.), 2% NaOCL, 0.2% detergent; 20 cycles.
[2]Polyacrylic acid (MW 4500).
[3]0 = no filming or spotting; 4 = entire surface covered with film or spots.

citrate and soda ash as the main builders. Many experts believe it is only a matter of time until nonphosphate formulations make their way into the U.S. market.

The development of a nonphosphate built automatic dish detergent necessitates the solution of many performance problems. Major deficiencies typically show up as excessive filming and/or spotting on glassware. Filming is usually caused by inorganic salt precipitation (calcium carbonate), and spotting usually results from inadequate removal or redeposition of protein-containing food soils. Shulman and Robertson [7] also show that fatty soils can also lead to significant filming and spotting under hard water conditions. These investigators evaluated sodium citrate/soda ash based formulations containing three different polymers to control hardness and food soil deposits, minimizing filming and spotting. They conducted multicycle dishwashing tests using margarine and dry milk as the soil source under hard and soft water conditions. Filming was caused primarily by fatty soil (margarine) deposition in hard water, whereas spotting was caused by the protein soil (dry milk) under both hard and soft water conditions (Table 5.8).

Conventional acrylic acid homopolymers and acrylic acid/maleic acid copolymers provide excellent control of filming and spotting caused by inorganic salts and fat soiling; however, they have little effect on protein soil deposition. Shulman and Robertson [7]

Table 5.8   Effect of Soil Components in Filming and Spotting under Soft and Hard Water Conditions

Soil type	Filming[1]	Spotting[1]	Water hardness (ppm)
Margarine (fat)	2.7	3.0	200
	0	0.1	20
Dry milk (protein)	0.7	3.5	200
	0	2.2	10

[1]0 = no filming or spotting; 4 = entire surface covered with film or spots.

showed that a hydrophobically modified maleic acid copolymer exhibits excellent control of spotting that is caused primarily by protein soiling. They recommend a total systems approach in developing nonphosphate formulations using sodium citrate and soda ash as the primary builders, and combinations of acrylic acid homopolymers, acrylic acid/maleic acid copolymers, and hydrophobically modified copolymers as cobuilders. A major deficiency of the hydrophobically modified maleic acid copolymers, however, is their lack of chlorine stability (see Section 5.4.7). See Chapter 7 for more details on dishwashing (Section 7.2).

5.3.3   I&I Laundry

Conventional I&I laundry detergents typically contain high levels of sodium hydroxide and sodium silicate, relying on high alkalinity to saponify greases and oils, which are the major soils encountered. Nonionic surfactants are added for the "suds"cycle. Historically, sodium carboxymethylcellulose (CMC) was used as the redeposition agent. Recent results reported by Schwartz [8] show that pAA/MA copolymers give better soil removal and antiredeposition performance than acrylic acid homopolymers on a variety of fabrics. The test materials were soiled with a broad range of oily and particulate soils and washed under I&I conditions in large-scale field trials.

Machines used in the I&I area vary from the small (35 lb) to the large (600 lb) washer/extractors. The bath-to-fabric ratio in these machines runs from about 4:1 (in 35 lb) to 1.8:1 (in 600 lb sizes). The low bath-to-fabric ratio in these machines means higher soil loadings, especially in the 600 lb machines. Significantly higher ionic strengths would be expected than in household machines, where bath-to-fabric ratios are about 25:1.

A tabulation of soil redeposition versus ionic strength shows that as ionic strength increased, polymer antiredeposition performance decreased (Table 5.9). The effect of ionic strength appears to have a greater negative influence on the performance of the acrylic acid homopolymer than on the acrylic acid/maleic acid copolymer. The higher sequestering capacity (Table 5.10) of the acrylic acid/maleic acid copolymer may explain its effectiveness under high ionic strength conditions. In 35 lb machines more water is used, and the ionic strength would be expected to be lower than in the 600 lb machines. Soil removal performance of an acrylic acid homopolymer is similar to that of the acrylic acid/maleic acid copolymer in 35 lb machines (Table 5.11) but poorer in the 600 lb machines (Table 5.12), where soil loadings are higher. In addition to improving soil removal, polymers can reduce soil redeposition under the high soil loadings typically encountered in I&I laundries (Table 5.13.)

Table 5.9   Soil Redeposition[1] Versus Background Ionic Strength

| Polymer | Molecular weight | Reflectance, Y, for three ionic strengths (NaCl)[2] | | |
		0.025 M	0.1 M	0.3 M
None		63	62	61
pAA	4,500	73	69	65
pAA/Maleic acid	40,000	77	73.5	72.5

[1]Dust/sebum on polyester/cotton (65:35); hardness, 75 ppm (softened water); 10 cycles.
[2]pH = 11.5.
Source: Ref. [8].

Table 5.10   Effect of Ionic Strength on Polymer/Calcium Sequestration Capacity

| (NaCl)[1] | Calcium sequestration capacity (mg CaCO$_3$/g sequestrant) | |
	pAA, MW 4,500	pAA/Maleic acid, MW 40,000
0.008 M	330	430
0.1	170	315

[1]pH 11.

Table 5.11   Soil Removal (Reflectance) Versus Stain Type in 35 lb Washer/Extractor

| Polymer[2] | Molecular weight | Soil removal[1] | |
		Blood	Chili
None		40	83
pAA	4,500	52	92
pAA/Maleic acid	40,000	53	93

[1]Alkalinity, 2500 ppm; Na$_2$O/SiO$_2$, 2:1; nonyl phenylethoxylate (EO)$_9$, 400 ppm; fabric, cotton.
[2]100 ppm.
Source: Ref. [8].

Table 5.12   Soil Removal (Reflectance) Versus Stain Type in 600 lb Washer/Extractor

| Polymer[2] | Molecular weight | Soil removal[1] | |
		Oily particulate	Blood
None		17	6
pAA	4,500	23	5
pAA/Maleic acid	40,000	32	26

[1]Alkalinity, 2500 ppm; Na$_2$O/SiO$_2$, 2:1; NPE$_9$, 800 ppm; fabric, polyester/cotton.
[2]100 ppm.
Source: Ref. [8].

Table 5.13   Soil Antiredeposition (Reflectance) Versus Stain Type

| Polymer | Molecular weight | Antiredeposition[1] | |
		Used cooking oil[2]	Used motor oil[3]
None		61	45.5
pAA	4,500	69	53
pAA/Maleic acid	40,000	76	58

[1]Alkalinity, 1200 ppm as Na_2O; NPE_9, 400 ppm; soft water.
[2]75 ppm polymer.
[3]100 ppm polymer.
Source: Ref. [8].

Since polymers are usually supplied as aqueous solutions, they can be readily incorporated into the highly alkaline I&I laundry formulations, forming easily pumpable products. CMC, however, is a solid and can be fairly difficult to incorporate into highly alkaline aqueous systems. Any insolubility can lead to pumping problems during use.

5.3.4   Food Process Cleaners

In the food process industry cleaning is generally done by the clean-in-place (CIP) method. Cleaning solution is circulated through the equipment and pipelines without dismantling the system. Liquid CIP cleaners are preferred for their ease of use in automated cleaning systems. Current formulations usually use potassium hydroxide in conjunction with sodium tripolyphosphate. KOH increases the solubility of STPP in the formulation, compared with the less expensive sodium hydroxide (NaOH). Replacing STPP with carboxylate polymers, such as acrylic acid homopolymers or acrylic acid/maleic acid copolymers, allows the formulator to use NaOH, thereby reducing formulation cost. Chlorine stability of these polymers is essential, since CIP cleaners contain hypochlorite as a sanitizing agent. Chelants, such as ethylenediamine tetraacetic acid (EDTA) and NTA, are not stable in hypochlorite and are not recommended for use in CIP formulations.

In general, the acrylic acid homopolymers give acceptable performance in water of moderate hardness (< 300 ppm). If hardness is greater than about 300 ppm, the acrylic acid/ maleic acid copolymer should provide adequate water hardness control as a result of its higher sequestering capacity relative to the acrylic acid homopolymer. pAA/MA copolymers have a higher binding capacity for calcium ions, compared to acrylic acid homopolymers (Table 5.14), and this may account for their enhanced soil removal performance. Milk soil removal tests comparing the performance of the two polymers are shown in Table 5.15.

5.3.5   Hard-Surface Cleaners

Hard-surface cleaners (window, tub and tile, floor, spray on–wipe off, metal and vehicle) are widely used throughout the home and industrial environment. Polymer usage in these areas is fairly new, compared with the laundry and dishwashing areas. In the consumer area, low levels of phosphate (either tripoly or pyro), are used. With the trend toward no-phosphate products, polymer usage is increasing. There are no current legislative pressures limiting the

Table 5.14   Calcium Binding Capacity[1]

Polymer	MW	Log k
pAA	4,500	2.2
pAA/Maleic acid	40,000	4.2

[1]Ionic background, 0.1 M KCl.

Table 5.15   Milk Soil Removal from Stainless Steel Panel[1]

Water hardness (ppm)	Surface appearance	
	pAA MW 4500	pAA/Maleic acid MW 40,000
200	Clean	Clean
350	1/3 surface soiled	Clean
400	1/2 surface soiled	Clean

[1]Detergent: 20% NaOH, 4% polymer, 1.1% sodium silicate, 2% NaOCl; 5 cycles; detergent level: 0.4%.

Table 5.16   Effect of Additives on Commercial Floor Cleaners

Cleaner[1]	Molecular weight	Cleaning efficiency (%)[2]
Home liquid (0.22%)		14.3
+ 2% TKPP		22.9
+ 2% pAA	4,500	18.8
+ 2% pAA/maleic acid	40,000	23.6
Home solid (0.63%)		27.6
+ 2% TKPP		36.0
+ 2% pAA	4,500	38.8
+ 2% pAA/maleic acid	40,000	41.9
Industrial liquid (0.69%)		45.9
+ 2%		59.8
+ 2% pAA	4,500	52.2
+ 2% pAA/maleic acid	40,000	60.1

[1]Percent solids basis.
[2]$(R_w - R_s)/(R_0 - R_s) \times 100$, where R_w = reflectance washed, R_s = reflectance soiled, R_0 = reflectance unsoiled. [Test Procedure D 4488-89-A5 of the American Society for Testing and Materials, Philadelphia.]

use of phosphates for these products; however, environmental concerns exist. In many hard-surface cleaning products little, if any, soda ash is used, and one of the key benefits of polymers—reducing filming or preventing encrustation—is not needed. What benefits can polymers contribute in these formulations? Results obtained by adding an acrylic acid homopolymer and an acrylic acid/maleic acid copolymer to various commercially available floor cleaning products (Table 5.16) show that polymers can make a visual improvement (about 4 reflectance units difference) in soil removal, in some instances, similar to that achieved by tetrapotassium pyrophosphate. See Chapter 7 for more details on hard-surface cleaning applications (Section 7.1).

5.4   Polymer Properties

Polymers are sold as aqueous solutions or dried powders. The various manufacturers characterize their products by the criteria examined in Sections 5.4.1–5.4.4.

5.4.1   Molecular Weight

The molecular weight (MW) of polymers used in the detergent area can range from about 1000 to 100,000. The choice of polymer depends on the needs of the formulator and the end-use properties desired. For very hard waters and high temperatures, polymer solubility (Fig. 5.7) may be critical, and polymers having lower molecular weights (1000–2000) may be needed. If bead strength during spray-drying is needed, polymers having a higher molecular weight (> 10,000) may be preferred. In general, the lower the molecular weight (within limits), the better the performance properties; the higher the molecular weight, the better the processing properties. However, processing problems usually can be handled with

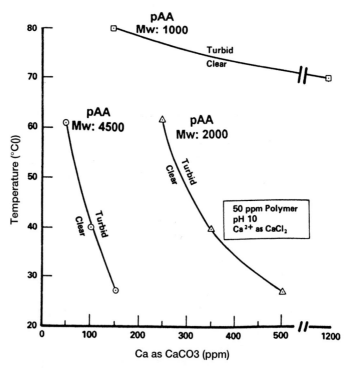

Figure 5.7   Polymer solubility as a function of water hardness and temperature.

equipment modifications. Today, most polymers are used to increase performance. Under U.S. conditions, polymers providing the best performance characteristics typically have molecular weights in the 4000–10,000 range.

Most polymer producers measure molecular weight using aqueous size exclusion chromatography [also known as gel permeation chromatography (GPC)]. Although this method is fairly easy to use, only relative molecular weights are obtained. The values obtained are dependent on the standards used to calibrate the equipment, and absolute standards are not currently available. End users need to be aware that reported polymer molecular weight can differ among suppliers. In the past few years a new matrix-assisted laser desorption time-of-flight mass spectrometry technique has been developed which shows promise of measuring molecular weights of polymers. This method would yield absolute values and could help standardize the industry.

5.4.2   Percent Solids

Most polymers are sold as aqueous solutions. The physical weight of polymer and nonvolatile inert material present in the product is referred to as percent solids. The two most common methods for measuring solids are heating in a forced-air oven (120–150 °C for 30–90 min) and heating in a microwave oven (5–15 min at maximum power). As the degree of neutralization increases, it becomes more difficult to remove water from the polymer. Significant differences in product solids can be obtained depending on the method used. Using a forced-air oven at 150 °C removes most water. The advantage of the microwave method is its much shorter drying time.

5.4.3   Degree of Neutralization (pH)

Polymers are supplied as acids, or partially to fully neutralized salts. The most commonly used cation is sodium, but potassium and ammonium salt versions are also available. Polymers prepared as acids tend to gel on standing as a result of hydrogen bonding. To minimize gelling, polymer solids are reduced, or the polymer can be partially or fully neutralized. In a fully neutralized polymer, about 25% of the product weight is sodium, whereas in a 20% neutralized polymer, only about 5% of a product weight is sodium.

5.4.4   Physical Form

Most polymers used are currently supplied as liquids. This is the least expensive form and is easily incorporated into aqueous formulations. Solid forms of the most widely used polymers are available either as spray-dried versions of the corresponding liquid product or as a premix of the polymer on a solid carrier such as soda ash or sodium sulfate. Dried, fully neutralized salts of the polymers are very hygroscopic and will readily absorb water and become sticky. Partially neutralized polymers are much less hygroscopic (Table 5.17).

Table 5.17   Water Absorption of an Acrylic Acid Homopolymer Versus Time at 25 °C and 70% Relative Humidity

	Weight gain (%)	
(min.)	20% Neutralized	100% Neutralized
0.5	0.5	4.4
1	1.0	5.0
2	1.6	7.7
4	3.8	13.6
6	5.6	18.8

5.4.5   Sequestration and Clay Dispersancy

Sodium tripolyphosphate is an excellent detergent builder for controlling water hardness in cleaning processes. A major advantage of STPP is its ability to form water-soluble complexes with ions that cause water hardness. One mole of tripolyphosphate ion $(P_3O_{10})^{5-}$ will combine with one mole of calcium ions (Ca^{2+}) to form the $Ca(P_3O_{10})^{3-}$ complex in aqueous solution. The anionic character of this complex also imparts dispersancy properties to the detergent formulation. Alternative builders such as soda ash, sodium citrate, tetrasodium pyrophosphate, and zeolite can remove hardness ions from solution but do not provide dispersancy properties. Polymers have sequestering capacity similar to that of STPP (Table 5.18). In general, however, they are used at relatively low concentrations (1–5% solids), compared with typical builder concentrations of 20–80%, and thus provide minimal hardness control. Sequestration capacity of polymers is directly proportional to their carboxyl content. Polymers containing maleic acid have higher sequestering capacity than acrylic acid homopolymers.

Carboxylate polymers contain a large number of carboxyl (–COO⁻) groups and can provide dispersancy and crystal growth inhibition at substoichiometric use levels. The chemical adsorption of an acrylic acid polymer onto a polar surface is illustrated in Fig. 5.8. The adsorption of the polymer increases the overall negative charge on the surface, increasing

Table 5.18   Polycarboxylate Sequestration and Dispersancy Properties

Polymer	Molecular weight	Calcium sequestration capacity (mg CaCO₃/g sequestrant)[1]		Kaolin dispersancy[2]
		pH 9	pH 11	NTU[3]
pAA	1,000	190	160	130
pAA	2,000	250	250	250
pAA	4,500	330	330	240
pAA	40,000	340	340	50
pAA/Maleic acid	70,000	380	430	130
STPP		320	320	

[1]Ionic strength, 0.008 M NaCl.
[2]400 ppm clay/ 50 ppm polymer; pH 10.
[3]NTU = nephelometric turbidity units.

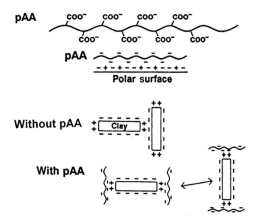

Figure 5.8   Polymer adsorption on polar surfaces.

dispersancy. Polymers of this type are used as dispersants for high solids slurries such as Crutcher mixes, kaolin, precipitated calcium carbonate, magnesium hydroxide, and other small particle size materials.

5.4.6   Caustic Solubility

Many I&I formulations for laundry and dishwashing applications contain high levels of alkalinity. In general, the acrylic acid homopolymers are compatible in formulations containing up to 35% alkaline solids. Some acrylic acid/maleic acid copolymers have fairly good alkali solubility, but some have rather poor solubility (Figs. 5.9 and 5.10).

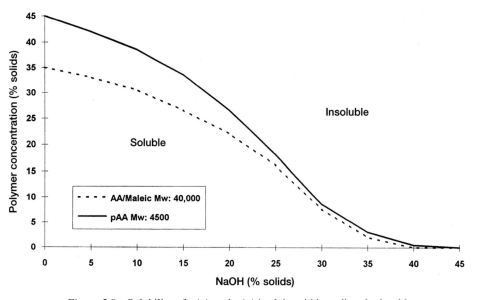

Figure 5.9   Solubility of pAA and pAA/maleic acid in sodium hydroxide.

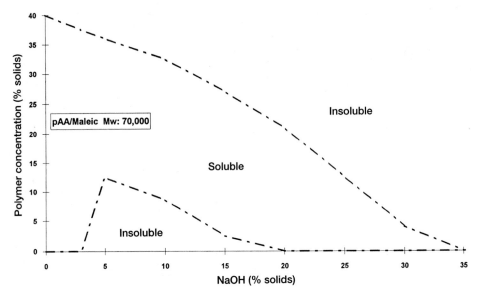

Figure 5.10   Solubility of a heavier pAA/maleic acid polymer in sodium hydroxide.

5.4.7   Hypochlorite Stability

Many of the existing home and I&I dishwashing as well as clean-in-place dairy formulations contain hypochlorite. The hypochlorite helps to reduce spotting and provides sanitation properties when used at a concentration of about 100 ppm. Polymers when used in these formulations need to be stable (i.e., they do not degrade themselves or increase the rate of hypochlorite loss). Many acrylic acid homopolymers and relatively few pAA/MA copolymers are stable enough for use in these formulations (Fig. 5.11).

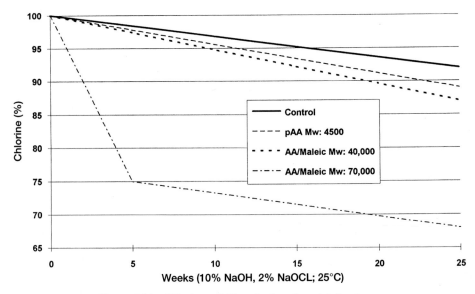

Figure 5.11   Chlorine stability of polycarboxylated polymers.

5.5   Thickeners/Stabilizers

Highly alkaline liquid formulations containing nonionic surfactants are commonly used in the I&I area. The nonionic surfactant in these formulations is typically above its cloud point and will phase separate. Polymers are added to these formulations to provide stabilization. Polymers commonly used are sodium carboxymethylcellulose, crosslinked polyacrylic acids, and high molecular weight carboxylic acid/ester alkali-soluble emulsion polymers. These products stabilize the formulation by increasing the viscosity of the system. In addition, some of these polymers can alter the rheology of the formulation to provide vertical cling for wall and grill cleaners.

   Thickeners can be classified as nonassociative and associative. The choice of thickener or stabilizer depends on the end-use needs of the formulator. All the thickeners described in Sections 5.5.1 and 5.5.2 are organic polymers and are susceptible to attack by strong oxidizing agents, such as hypochlorite. Some impart a Newtonian rheology while others can be highly shear thinning. The nonassociative types form products with minimal viscosity drift, whereas the associative types may exhibit appreciable viscosity drift, especially after the first few hours of product production. The thicker the product, the longer it takes for it to reach its final equilibrium viscosity.

5.5.1   Nonassociative

Nonassociative thickeners, which exert their action on the water phase only, fall into the following categories:

Naturally derived
     Cellulose ethers
     Other polysaccharides
Synthetic polymers
     Alkali soluble
     Alkali swellable
     Nonionic

5.5.2   Associative

Associative thickeners, which interact with the dispersed phase, are grouped as follows:

Naturally derived polymers
     Hydrophobically modified cellulose ethers
Synthetic polymers
     Nonionic urethanes
     Hydrophobically modified, alkali soluble

5.5.3   Physical Form

Commercial thickeners and stabilizers are sold in both solid and liquid forms. Care must be used to prevent clumping when the solid products are added to aqueous solutions.

References

1. Holland, R., Household Pers. Prod. Ind., *28*(4), 100 (1991).
2. Jaeger, H.U., Denzinger W., Tenside surfactants, Detergents, *28*(6), 428–433 (1991).
3. Burns, M.E., Nicol, C.H., U.S. Patent 4,174,305, Procter & Gamble Company, 1979.
4. Spadini, G.L., Larrabee, A.L., Liu, D.K., U.S. Patent 4,490,271, Procter & Gamble Company, 1984.
5. Chem. Mark. Rep., *243* (1), 36 (1993).
6. Jones, C.E., Witiak, D., Household Pers. Prod. Ind., *27*(8), 64 (1990).
7. Shulman, J.E., Robertson, M.S., Soap Cosmet. Chem., *68*(11), 46 (1992).
8. Schwartz, C., Household Pers. Prod. Ind., *29*(12), 84 (1992).

CHAPTER 6

Laundry Products

Thomas Christ, William W. Morgenthaler, and Frank Pacholec

This chapter concentrates on the household detergent industry as it exists today, with some indications of trends that presage future directions for laundry compounding. The basic ingredients include surfactants and builders and, to enhance cleaning, functional additives such as bleaches, optical brighteners, and antiredeposition agents. How and why these ingredients are formulated and manufactured as they are, and how the resulting products are then presented to the market, comprise the main thrust of this chapter.

Market needs demand the complete evaluation of products to ensure consumer satisfaction. Today's market also demands that the supplier demonstrate environmental consciousness, from the packaging of the product to its eventual treatment as waste. Consumer and environmental forces are bringing about marked changes in the field, generally in the direction of concentrated products, both solid and liquid, to save on packaging and shelf space. Manufacturing methods and formulation practices are evolving rapidly, away from traditional spray-dried powder toward agglomerated, denser product and to stabilized concentrated liquids.

6.1   Introduction

Since the introduction of synthetic detergent formulations in the mid-1940s, the industry has steadily grown into a multi-billion-dollar business. In 1992 the heavy-duty laundry detergent market was expected to reach $3.85 billion in total sales [1].

This chapter provides a practical overview of laundry detergent formulating as it is practiced in North America. An attempt has been made to cover the basics pertaining to a number of different areas within this broad topic, to give the reader an overall perspective of the elements involved in manufacturing and marketing laundry detergent products. References have been selected to provide current, practical information for readers interested in exploring a specific topic in greater depth.

6.2   Detergent Ingredients

Detergents used in home laundry and commercial applications are complex formulations consisting of a myriad of chemical ingredients, which can be classified into three major

Thomas Christ, William W. Morgenthaler, and Frank Pacholec, Monsanto Company, 800 N. Lindbergh Blvd., St. Louis, Missouri 63167, U.S.A.

Table 6.1   Typical Ingredients Found in Laundry Detergents

Ingredients group	Dry products	Liquid products
Surfactants		
Anionic	Linear alkylbenzenesulfonates (LAS)	LAS, AEOS
	Alkane sulfonates (AS)	
	Fatty alcohol sulfates (FAS)	
	Alcohol ethoxysulfates (AEOS)	
	Fatty acid soap	
Nonionic	Alcohol ethoxylates (AEO)	AEO
Builders		
No-P	Sodium carbonate	Sodium citrate
	Zeolite A	
Phosphate type	Sodium tripolyphosphate (STPP)	
Auxiliary		
Bleaches	Sodium perborate	
Enzymes	Protease, lipase	Protease, lipase
Optical brighteners	Pyrazolines	Stilbenes
	Stilbenes	
Cobuilders	Polycarboxylate	Polycarboxylate
	Sodium citrate	
Anti-redeposition	Sodium carboxy methyl cellulose	
Corrosion inhibitors	Sodium silicate	
Fillers	Sodium sulfate	Water
Hydrotropes	Sodium xylene sulfonate	Sodium xylene sulfonate
	Sodium toluene sulfonate	Sodium toluene sulfonate
Solvent		Short-chain alcohols

groups: surfactants, builders, and auxiliary agents. Though these ingredients contribute specific functions to the cleaning process, synergistic effects may operate between ingredients. Thus a certain amount of "artistry" and experience is required to develop a high-performing formulation. This section presents an overview and brief discussion of each ingredient category. A more thorough review of ingredients is available in Chapters 3, 4, and 5. Additional information on ingredients, trends, and the general detergency market can be found in the literature [2–10]. Table 6.1 provides a broad summary of typical ingredients found in North American home laundry detergent formulations.

6.2.1   Surfactants

Surfactants, the most important group of detergent components, have been written about extensively [2,11–16]. In a detergent, the surfactant(s) provide cleaning action. The primary purpose of a surfactant is to facilitate the solubilization of species that exhibit poor solubility in the cleansing medium (generally water). Examination of the molecular structure of surfactants reveals that they generally consist of a polar or hydrophilic (water-loving) portion, and a nonpolar or hydrophobic (water-hating) section. It is the hydrophobic moiety that has an attraction for nonpolar species (fats, oils, greases, etc.) and enables the solubilization of these species in water. In addition to soil removal, a surfactant suited for detergent use is expected to possess several other characteristics, such as the following [17]:

* Acceptable environmental behavior
* Low sensitivity to water hardness
* Dispersion properties
* Soil antiredeposition
* Adequate solubility
* Wetting capability
* Acceptable foam characteristics
* Neutral smell and low color
* Storage stability
* Specific adsorption
* Very low human toxicity
* Plentiful raw material source
* Reasonable cost

Surfactants can be classifed into four main types, based on the nature of the chain-carrying part of their molecular structure when dissolved in water: anionic, cationic, nonionic, and amphoteric. These are discussed here in a general fashion; for details, see Chapter 3.

Anionic surfactants have a negatively charged ion (anion) incorporated into the chain-carrying portion of the molecule. They are the main surface active agents used worldwide in today's laundry detergents. Anionics possess a number of outstanding properties, including soil removal and dispersion power [6]. They are generally high sudsing. Examples of anionic surfactants include linear alkylbenzenesulfonates (LAS), fatty alcohol sulfates (FAS), alcohol ethoxysulfates (AEOS), secondary alkane sulfonates (SAS), and α-olefinsulfonates (AOS).

Nonionic surfactants do not dissociate to produce ions when dissolved in water. Nonionics can be used alone or in combination with anionic or cationic surfactants in detergents. The principal representatives of this class of surfactants are the ethoxylated alcohols (AEO), which can be derived from natural fatty alcohols or oxoalcohols of petrochemical origin. Nonionic surfactants tend to be low foaming, and they exhibit excellent detergency for removal of oily and greasy soils, especially from synthetic fabric. Other examples of nonionics are alkylphenol poly(ethylene glycol) ethers (APE) and fatty acid alkanolamides (FAA).

Cationic surfactants, which have a positively charged ionic group in the chain-carrying portion of the molecule, are not generally used for soil removal specifically. Though mixtures of equivalent amounts of anionic surfactant and cationic surfactants are unabsorbed and exhibit no detergency, small amounts of some cationics added to anionic surfactant based formulations can enhance detergency performance [2]. Cationic surfactants are more often used in fabric softener and antistatic applications, sometimes in conjunction with nonionics. The most commonly used class of cationics are the quaternary ammonium compounds.

Amphoteric surfactants contain both anionic and cationic groups. The form existing in aqueous solution depends on the relative acidity or basicity (pH). Though this class of compounds exhibits some excellent detergency properties, amphoterics are not commonly used in detergents because of high cost; they do find use in shampoos and other personal care products, however. Examples of amphoteric surfactants include imidazoline derivatives, alkylbetaines, and alkylsulfobetaines.

6.2.2   Builders

Detergent builders are used in essentially all home laundry powders as well as in many home laundry liquid detergents. They improve the cleaning efficiency of surfactants, primarily by inactivating water hardness ions (calcium and magnesium) that can interfere with surfactant performance, thus "building" the effectiveness of the detergent formulation. Depending on the specific builder, this improvement may be accomplished through chelation or sequestration (complexation), ion exchange, or precipitation.

The use of builder systems that act solely by *precipitation* (generally based on alkalies or monophosphates) is outdated in North America. However, systems based on sodium carbonate, and containing polymeric compounds and/or mixtures of other builders, continue to be popular.

Builders of the *complexing agent* type can be divided into two types: those that contain phosphorus and those that do not (no-P). These compounds act by chelating water hardness ions, to keep them in solution, but disabling them from interfering with surfactant performance. Since the late 1960s, the use of phosphate builders, or in the case of home laundry detergents sodium tripolyphosphate (STPP), has been increasingly criticized because of alleged contribution to early eutrophication of bodies of water. While significant research effort has been dedicated to the development of no-P builders, few have been commercialized. The first compound to be introduced as a result of this effort (early 1970s) was sodium nitrilotriacetate (NTA). Though NTA is the builder of choice in Canada and is used in parts of Europe in home laundry applications, it is not used in the United States because of possible carcinogenicity. Such allegations, however, have been heavily debated by scientists all over the world.

Zeolites (sodium aluminosilicates) are insoluble in water and act via an ion-exchange mechanism (vs. complexation). Zeolites (in particular, zeolite 4A) are finding increased use as STPP is phased out and will play a key role in the home laundry builder systems of the future. No-P builder systems currently used in home laundry applications are comprised of a mixture of builders and cobuilders, including sodium carbonate, sodium citrate, silicates, polycarboxyates, and zeolites. Additional information on builders may be found in the literature [6,7,18–22] and in Chapter 2, Mechanisms of Soil Removal.

6.2.3   Auxiliary Ingredients

As discussed above, surfactants and builders are the two major components of modern detergent formulations. However, a variety of other ingredients is added to formulations, either to perform specific functions (e.g., bleach ingredients, optical brighteners, enzymes, antiredeposition agents, foam regulators) or to enhance product appeal to the consumer (dyes, perfumes).

6.2.3.1   Bleach Ingredients

Bleach ingredients are compounds that can clean, whiten, and brighten fabrics and remove stains [2]. They may be included in the actual detergent formulation or added with the detergent in the laundering process.

6.2.3.1.1   Types of Bleach

The most widely used bleach for home laundry applications in North America is liquid chlorine bleach. It is usually introduced to the laundering process separately from the detergent as a sodium hypochlorite solution. Sodium hypochlorite is a powerful bleaching agent and it is well suited for the short wash cycles and lower wash temperature conditions prevalent in North America.

Oxygen-based (peroxide) bleaches are usually powders containing sodium perborate, which serves as a source of hydrogen peroxide when dissolved in an alkaline medium. It is the peroxide species that leads to the bleaching action. Perborates may be included as part of a separate dry (powder) bleach product, or as part of a detergent formulation, where they exhibit excellent shelf life in the absence of trace metals. Several pertinent publications on bleaching may serve readers' further interest [2,23–25]. Chapter 4 gives details on bleach systems used today in laundry detergents (Section 4.1).

6.2.3.1.2   Bleach Activators

Bleach activators are commonly used to achieve satisfactory bleaching with sodium perborate at reduced wash temperatures (< 60 °C or 140 °F) [1]. These acylating agents react with hydrogen peroxide to generate other species that exhibit bleaching action at lower temperatures. A variety of different activators may need to be employed, depending on the range of wash temperatures under consideration.

6.2.3.1.3   Bleach Stabilizers

Some trace metals can catalyze the release of oxygen from peroxide bleach systems, leading to decreased bleach effectiveness and possible damage to fabric. Therefore complexing agents such as organo phosphonates or aminopolycarboxylates may be added to a detergent formulation to inactivate the action of trace metals on the bleach.

6.2.3.2   *Optical Brighteners*

The purpose of optical brighteners (also known as fluorescent whitening agents) is to visually enhance fabric whiteness during laundering. Properly washed laundry, even when "completely clean," has a slight yellow tinge. Optical brighteners such as stilbenes and pyrazolines create a visual whitening effect when exposed to near-ultraviolet radiation by virtue of fluorescence [26]. Several useful reviews have been written in this area [27–30], and Chapter 4 presents a detailed review of this topic (Section 4.2).

6.2.3.3   *Enzymes*

Some protein stains and fat-containing soils are very resistant to detergents. Proteolitic (protein-cleaving) enzymes or lipases (fat-cleaving) enzymes can be added to a formulation to facilitate the removal of such stains, which may be derived from foodstuffs (e.g., milk, egg), plants, or body fluids (e.g., blood). Many experts feel that enzymes will play an increasingly important role in future formulations as washing conditions become milder in response to energy and environmental considerations in the form of lower wash temperatures and use of nonphosphate detergents, respectively. More information is available in several articles published on this topic [31–40].

6.2.3.4   Antiredeposition Agents

Soil removed in the washing process is normally finely dispersed. Through judicious choice of surfactants and builders, this dispersed soil is generally not redeposited on laundry items. If this goal is not achieved, laundry items may look gray after repeated washings. Specific compounds such as carboxymethylcellulose (CMC), polyacrylates, other polymers, and/or derivatives thereof may be added to a formulation to aid in antiredeposition by binding irreversibly to certain textile fibers (cellulose-containing in the case of CMC) and soil particles. See Chapter 5 for a complete discussion of polymers used in laundry products. Chapter 2 presents the mechanisms of soil removal and suspension.

6.2.3.5   Foam Regulators

Foam boosters are compounds that provide a degree of foaming in the wash that can be appealing to consumers. Fatty acids, fatty acid amides, fatty acid alkanolamides, betaines, soap, sulfobetaines, and amine oxides have been used for this purpose [1]. In some washing situations (or market niches), it may be desirable to minimize foaming by employing a foam regulator. Nonionic surfactants having a hydrophile/lipophile balance (HLB) of less than nine or aliphatic alcohols, C_8 or higher, can reduce foam in the wash cycle.

6.2.3.6   Corrosion Inhibitors

A corrosion inhibitor is a compound that protects the surfaces of washing machines and/or any metallic components of clothing (zippers, clasps, etc.). Sodium silicate is the main corrosion inhibitor used in home laundry detergents. When added to such a formulation, it will protect washer metal parts and finishes, especially porcelain enamel.

6.2.3.7   Dyes

Dyes are sometimes used to color part or all of a detergent formulation. Some products have appeared on the market as white powders containing colored granules, while others have been entirely colored. The main reason for putting color into formulations is to provide aesthetic value and to distinguish them in the marketplace, although blue colorants specifically provide bluing action on fabrics. Any dye added should have good storage stability and should not adversely affect the fibers of laundry items.

6.2.3.8   Perfumes

Perfumes (or fragrances) were first added to detergent formulations in the 1950s. A given perfume may affect the odor of the product itself, the washing water, the headspace above it, or the laundry items (or a combination thereof). Perfumes are generally present in small concentrations and tend to be complex mixtures.

6.2.3.9   Fillers/Formulation Aids

Fillers may be used to improve the flow, solubility, caking, dusting properties, or consumer appeal of powdered detergent products. The most commonly used filler for solid-type products is sodium sulfate. Dye stability on fabrics, an important property before the introduction of modern dyes and synthetic fabrics, is enhanced by high salt levels.

In the case of liquids, formulation aids are added to ensure their solubility and stability. Compounds employed for this purpose include polyglycol ethers, low molecular weight alcohols, and short-chain alkylbenzenesulfonates [2]. The recent trend toward concentrated powder and liquid detergent forms has led to a decrease in the use of fillers.

6.3   Formulation Considerations

The decade of the 1980s and the early 1990s have seen unprecedented growth in development and introduction of new products into the household laundry detergents market. Manufacturers of major branded products have introduced product line extensions through product modifications, as well as total formulation changes and packaging, to meet convenience/performance-oriented criteria for household detergent products (e.g., premeasured dosing along with concentrated powders and liquids).

The majority of changes in laundry product offerings are driven by manufacturers' response to some of the following concerns: safety and environmental issues, demographics of the American consumer, and individual consumer tastes and preferences [6,41]. The trend toward multifunctional and convenience-type laundry products drives researchers toward new concepts and practices in manufacturing, and toward new ingredients (surfactants, enzymes, bleaches, etc.), designed to deliver product performance as required for specific market niches.

New formulation considerations are needed to produce these multifunctional detergents and new product forms. As a result, issues of ingredient compatibility, liquid stability, solubility of powders, product integrity, and increased manufacturing cost must be considered. Solving these formulating problems usually results in an increased number of ingredients in formulations required to meet multifunctional performance criteria.

6.3.1   Product Types

Prior to the 1980s, detergent formulators had fewer options in developing laundry detergent formulations. Spray-drying of detergents was the preferred method of production, and powders were the main product used by consumers. Laundry powders of this era were usually based on a single surfactant–builder system. Sodium tripolyphosphate was the builder of

Table 6.2   Examples of Modern Home Laundry Products: Regular
Versus Concentrated Powder Phosphate Containing Formulations

Ingredients	Composition (wt %)	
	Regular	Concentrated
Surfactants		
Sodium alkylbenzenesulfonate	14	17
Sodium alcohol sulfate	3	7
Alcohol ethoxylate	5	1
Builders		
Sodium tripolyphosphate	24	24
Zeolite A	—	12
Sodium carbonate	13	23
Sodium silicate (solids)	7	3
Additives		
Sodium perborate[1]	5	5
Sodium sulfate	20	8
Enzymes	0.5	0.5
Water	5	4
Properties		
Density, g/cm^3	0.33	0.54
Use level, cm^3	240	120

[1] Some detergents use a perborate activator, *acylating agent,* to increase
bleach performance during the wash process.

choice, in conjunction with either a single anionic (linear alkylbenzenesulfonate) or nonionic (alcohol ethoxylate) surfactant. These powder products were easy to classify as either "anionic" or "nonionic."

Today there are many more options to consider in classifying product types: powders (with or without phosphate), regular versus concentrated, and liquids and multifunctionals. Laundry products formulated with mixtures of several builders and surfactant blends, along with special additive functional ingredients, are broadly classified as either powders or liquids. Examples of the ingredient complexity of powder laundry formulations are shown in Table 6.2.

The formulations shown in Table 6.2 also illustrate how a manufacturer can use similar ingredients at varying levels to deliver comparable detergent active use levels to the consumer. For example, at one-cup and one-half-cup levels, respectively, (in a 17-gallon washer), both these products will deliver equivalent (ca. 260 ppm) use levels of surfactant. Thus by increasing product density and total surfactant content, an equivalent amount of active ingredients is delivered at half the consumer use level (for the concentrated formulation). Therefore, targeted consumer end use of detergent products will have significant impact on processing, ingredient compatibility, packaging, and final product form.

6.3.1.1   Market Niche

In heavy-duty laundry detergents, the largest market growth segment in the 1980s consisted of liquids, marketed as "convenience" products to replace powders. Liquid convenience products offer advantages such as easier pretreatment of stains prior to washing, complete solubility in cold water, easier dispensing from the package, and no caking due to moisture when the product is stored [42,43]. Reducing energy use by lowering wash temperatures and decreasing use of phosphates in powders has significantly enhanced consumer acceptance of liquid formulations. Since 1984, the liquids' market share of total laundry detergents has increased from 20% to slightly greater than 50% in areas of the United States where phosphates are banned. In the last few years, manufacturers have introduced liquid brand name extensions for almost every major brand powder product. There are two types of liquid available to consumers, built and nonbuilt. Both types are isotropic (clear single-phase) products.

Liquids are easily measured; they dissolve in cold water, and they can be applied directly on stains as a pretreatment. Laundry liquids that are designed to clean and perform additional functions as antistats (static "cling" reducers), softeners, soil release agents, and specific eliminators of certain stains are known as multifunctional products [44]. Consumers who read labels will find these products being advertised or aimed at specific laundry issues for market-niched products. Each manufacturer has its own marketing approach to achieving economical production, effective pricing, and aesthetic properties for liquid detergents. A common characteristic of all liquid formulations is the heavy dependence on surfactant(s) choice to provide acceptable cleaning performance. Table 6.3 lists some typical liquid formulations, which are classified as built or nonbuilt depending on whether a builder is part of the formulation.

Performing the same calculation described earlier for powders reveals that at one-half-cup use, built liquids deliver about 540 ppm surfactant, versus nonbuilt liquids at 230 ppm surfactant use. Consideration of product performance related to recommended end-use levels of active ingredients is necessary when formulating and optimizing cost of ingredients for these market-niched products.

Table 6.3   Examples of Modern Home Laundry Liquid Formulations

	Composition (wt %)	
Ingredients	Regular	Concentrated
Surfactants		
Sodium alkylbenzenesulfonate	15	
Sodium ether sulfate	7	12
Alcohol ethoxylate	5	15
Builders		
Sodium citrate	10	
Additives		
Hydrotrope(s)	4	4
Solvent	5	5
Enzymes	0.5	0.5
Water	Balance	Balance
Properties		
Density, g/cm^3	1.1	1.0
Use level, cm^3	120	60

6.3.1.2   Types of Soils and Stains

There are many ways to categorize major laundry soils. Generally, consumers consider oily and particulate soils to be of household origin, whereas food and protein soils are classified as stains. That is, routine household soils like oily/particulate or collar and cuff soil are cleaned by heavy-duty detergents; food and protein soils/stains occur less frequently and require the consumer to clean, by special case or attention, as a problem soil. Table 6.4, which lists soils and stains, also gives typical ingredients used to optimize detergent formulations for cleaning. From a detergent formulation and ingredient selection standpoint, manufacturers optimize ingredient mixtures to function on targeted soils/stains, which they then advertise on product labels.

Oily and particulate soils are removed by surfactants and builders via specific mechanisms found in the wash process [44–46]. Surfactants are selected to optimize their functional work of adsorption at air–water and solid–liquid interfaces present during washing. In effect, surfactants do the work of wetting and roll-up of oily soils on fiber substrates to suspend and emulsify oily soils during the wash process [45,47,48]. Builders are present to enhance the ability of surfactants to remove soils by chelating (binding) potential interfering polyvalent metal ions (e.g., calcium or magnesium) present in wash water. Soils based on food and protein/starchy materials (i.e., natural dyes found in fruits, blood, tea, and wine) are removed

Table 6.4   Soils, Stains, and Ingredients for Optimizing Detergent Performance

Soils	Ingredients
Oily soil	Surfactants
Human sebum	
Animal fat	
Vegetable oil	
Mineral oil	
Waxes	
Food	Bleaches
Fruits	
Vegetables	
Wine	
Coffee	
Tea	
Particulates	Builders
Clays	
Metal oxides	
Humus	
Carbon	
Proteins/carbohydrates	Enzymes
Blood	
Egg	
Milk	
Grass	
Starches	

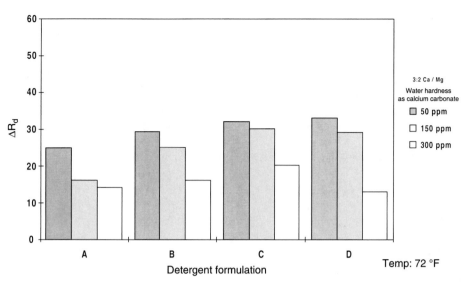

Figure 6.1   Detergency effects on polyester/sebum soils versus water hardness on four consumer detergent products.

by oxidation, while enzymes cleave protein-bound soils by decomposition. All these processes are influenced by thermodynamic and kinetic factors of ingredient interactions with soils/fabrics and are related to product end use by consumers in the wash process. It is important to realize not only that detergent ingredients interact and/or react with each other, but mixed soils can interact on fiber substrates as well [49,50]. See Chapter 2.

The soil removal process is thought of as having primary and secondary properties in the wash process. Typically, in detergent performance testing of a formulation's soil removal properties, primary soil removal can be measured by single-cycle test methods. Figures 6.1 through 6.3 are examples of testing results obtained by single-cycle Terg-O-Tometer tests to compare performance of four detergents by the measured laboratory performance scale, ΔR_d. There are varying cleaning responses due to each fabric/soil combination for each detergent. These are normal performance responses showing fabric–soil interactions due to formulation ingredients under a given set of test conditions. The secondary properties, such as graying or yellowing of fabric, require multicycle testing to measure the effectiveness of ingredients on soil antiredeposition and encrustation of fabrics. Secondary properties are related more to builders and carboxymethycellulose or polymer functions as threshold agents in detergent formulations.

No single type of soil or stain can predict overall detergency performance. Even using a variety of soils and stains under controlled conditions, performance testing of detergent formulations should be considered to be a guide for ingredient selection, not a substitute for consumer experiences.

6.3.1.3   Wash Variables

Soil removal during washing is controlled by three energy functions. Thermal, mechanical, and chemical energy balance the thermodynamic contributions of the soil removal process.

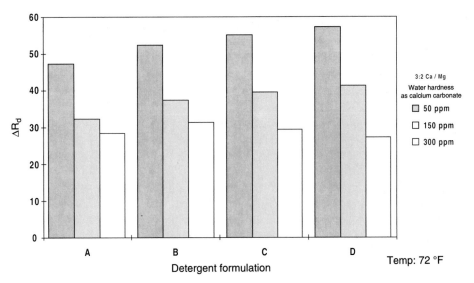

Figure 6.2   Detergency effects on cotton/sebum soils versus water hardness on four consumer detergent products.

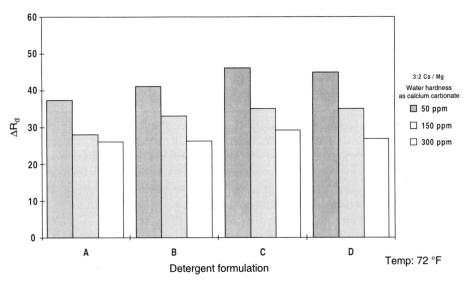

Figure 6.3   Detergency effects on cotton/clay soils versus water hardness on four consumer detergent products.

As wash temperatures are decreased, an increase in either mechanical agitation or chemical energy is required to obtain acceptable soil removal. To guide manufacturers and researchers in product development to meet an identified consumer market need, laboratory test methods are developed to simulate as many consumer experiences or variables as possible. The following variables usually are considered in the initial stages of the development of a laundry product.

Intrinsic factors (basic, interactive, thermodynamic). Type and amount of:
- Soils
- Fabrics
- Water hardness
- Surfactants
- Builders
- Additives (fluorescers, enzymes, bleaches, antiredeposition agents)

Extrinsic factors (external, kinetic):
- pH
- Temperature
- Time
- Fabric/water ratio
- Mechanical agitation
- Solubility

Aesthetic factors:
- Foam in wash/rinse
- Odor, color, form, etc.

These testing factors are used as a guide for evaluating soil and stain removal performance of home laundry products under home laundering conditions. Performance results obtained from laboratory testing utilizing these washing factors guide manufacturers in selection, preparation, and application of laundry products, as well as in targeting specific market niches.

Product testing temperatures for U.S. wash conditions can range from 5 to 60 °C (40 to 140 °F), but more practical temperatures are 24–49 °C (75–120 °F). Water hardness ranges from 0 to 300 ppm (as $CaCO_3$) with 2:1 or 3:2 Ca/Mg mole ratios. With so many variables to consider, the goal of laboratory product testing is to define the ultimate performance differences of ingredients or formulations across a number of fixed and variable washing conditions, which show measurable soil and stain removal performance.

A study employing experimental design (e.g., Placket–Burman) is the most rapid, efficient means of identifying key variables and determining specific effects on product performance. More elaborate statistical design methodology can be employed when testing for formulation component interactions that may produce nonlinear response to changes in wash conditions.

6.3.1.4   Fabric Types

Relative performance of household detergents will be influenced markedly by the nature of the fabric substrate. Detergency testing usually includes more than one type of fabric–soil combination. Performance results should be compared by individual fabric–soil types under a fixed set of testing conditions. Figures 6.1 and 6.2 plot performance comparisons of detergency results, showing soil removal differences using the same soil on two different fabric substrates. As can be seen by measured soil removal reflectance data using the same detergency test condtions, ΔR_a, sebum soil is much more difficult to remove from polyester cotton than from cotton fabric.

The most common fabric types found in U.S. household laundry applications are 100% cotton, polyester/cotton blends (with permanent press finish), 100% polyester, and nylon [2,51].

All fabrics to be tested should be prewashed prior to application of soils and stains, to remove residues of mill finishing agents that may influence soil removal results. Fabric construction is also important from the standpoint of staining/wicking characteristics of test soils. For example, bed linens will give oil wicking patterns different from those of shirting materials. It is also necessary to keep the orientation of fabric grain (warp and weave) consistent throughout soiling to achieve consistent results by visual means and/or instrumental reflectance.

6.3.2   Watch-Outs: Safety

In the initial stages of laundry detergent product development, all aspects of product safety must be of highest priority. One cannot assume that consumers will adhere to warning labels or safety precautions. Every possible step should be taken to maximize human health and environmental safety considerations for normal use and anticipated misuse of the product. Thus, ingredient selection must be based on low risk to humans, animals, and the environment, as well as the specific detergent action desired for the formulated product. Complete human and environmental risk assessments must be conducted on raw materials. Also required is safety testing for skin sensitivity, eye contact, inhalation, and ingestion, along with longer term testing for carcinogenicity and mutagenicity. Since most laundry products eventually end up in wastewater of one type or another, tests on ingredients are also conducted for long-term ecological effects.

6.4   Detergent Processing

6.4.1   Introduction

As described previously, laundry detergent products are composed of a variety of inorganic and organic components, each with specific contributions to either the cleaning process, the properties of the final product form, or detergent processing. The most cost-effective ingredients have a role in all three of these key elements of detergent production and usage, whereas some contribute in only one category.

The goal of detergent processing is to convert several raw materials into aesthetically pleasing finished products that have the desired physical properties and cleaning characteristics. Some of the key finished product characteristics for dry detergent products are uniformity, flowability, absence of caking, minimal dusting, bulk density, and solubility. Key characteristics of liquid detergents include homogeneity, storage stability, and solubility.

6.4.2   Powder Detergent Processing

A variety of processing techniques are suitable for producing detergent products. They range from simple batch-mixing operations through the more sophisticated continuous agglomeration and spray-drying techniques. More recently, combinations of processing techniques have evolved to produce the increasingly popular compact, or concentrated, dry detergent

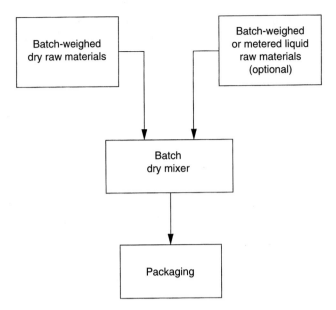

Figure 6.4   Block diagram of dry mixing operation.

product form. The literature contains several reviews of detergent processing techniques [2,52–57].

6.4.2.1   Dry Mixing

Dry mixing is the simplest processing technique for manufacturing powder detergent products. Typically, dry mixing is a batch operation utilizing various types of paddle, ribbon, or tumbling blenders. Liquid formulation ingredients, such as nonionic surfactant, are typically sprayed into the mixer and absorbed by the granular or powder components. Other liquids are reacted "in situ" on the powder bed to form a surfactant salt, such as the dry neutralization of dodecylbenzenesulfonic acid. Figure 6.4 is a block diagram of a simple batch dry mixing operation.

Advantages of dry mixing include low capital investment and high flexibility. The same equipment can be readily used to make a variety of products. However, there are a number of disadvantages associated with this processing technique. Batch operation is not well suited for large volume (multi-million-pound) production of detergent powders. Dry blending is almost totally dependent on raw material properties for density and sizing control. In addition, there are limitations in raw material selection and the quantity of liquid components that can be incorporated in the finished products. In many cases, these limitations result in cost penalties. Finally, the aesthetic properties (e.g., uniformity and particle sizing) of products made by this technique do not compare favorably with those obtained via alternative processing methods. In some cases the use properties, such as rate of solubility, are deficient.

Thus, although dry mixing is one of the three main processing options for a variety of cleaning products, its use for consumer laundry detergents is limited to relatively low volume, budget-type products. This technique is used extensively for making a variety of laundry and specialty products for the industrial and institutional (I&I) cleaning market because of the ease of making different formulations.

6.4.2.2   Spray-Drying

Spray-drying has long been established as the preeminent processing technique for dry consumer laundry detergent products. Today, this processing method is practiced worldwide to generate high-volume detergent products with an excellent balance of physical, aesthetic, and performance characteristics.

Spray-drying technology, like the detergent industry, has grown, developed, and become more sophisticated over the past 50 years. An informative historical perspective on the evolution of detergent processing with emphasis on spray-drying is provided by Jakobi and Lohr [2].

Although some truly continuous spray-drying plants are operational, the majority of detergent spray-drying plants are run in a batch–continuous operational mode. Dry and liquid raw materials are metered, in a specific order of addition, into a batch mixing vessel called a crutcher. The resulting detergent slurry typically has a high solids content to minimize the amount of water that must be evaporated in the subsequent spray-drying operation. Crutcher slurries are transferred to a stirred storage tank, becoming the feed for the continuous part of the process. Slurry is transferred through a series of transfer pumps, sieves, wet mills, and high-pressure pumps to a set of spray nozzles in the top of the spray tower. The slurry is atomized through the spray nozzles, forming spherical droplets that fall through a stream of countercurrent hot air. Upon contacting the hot air, the outer surface of the spherical droplet quickly dries. The remainder of the water evaporates from inside the bead by blowing a "hole" in the bead as it falls through the tower.

The spray-dried beads are removed from the tower and cooled during subsequent material handling steps entailing equipment such as an air lift or a belt conveyor. A variety of in-line spraying and/or mixing techniques is employed for adding heat-sensitive materials (perfume, enzymes, bleach components, etc.) to complete the processing to finished product ready for packaging. Figure 6.5 is a flow diagram for a typical countercurrent detergent spray-drying plant.

The variables that determine the physical properties of the finished spray-dried detergent include formulation, slurry composition and solids level, spray nozzle selection, atomization pressure, air temperature, and throughput. These and other variables must be optimized for each detergent formulation but are generally controllable within fairly specific limits to produce consistent, uniform products. Typical spray-dried products have bulk densities in the range of 0.25–0.5 g/cm^3, although there are examples of spray-dried products with bulk density up to 0.65 g/cm^3.

The majority of worldwide powder detergent products are produced by spray-drying. This processing technique lends itself to high-volume production with great flexibility in raw material product forms. The products are uniform in particle size and density, have good to excellent solubility characteristics and, when processed properly, have excellent flow properties. Spray-drying has the highest energy requirements among typical detergent processing techniques. Good dust collection systems are essential because of the large volumes of air moving through the tower.

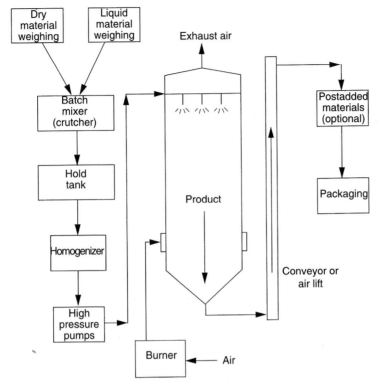

Figure 6.5   Block diagram of a countercurrent detergent spray-drying operation.

6.4.2.3   Agglomeration

Agglomeration processing of detergent products has historically been associated with production of automatic dishwashing compounds. Use of this processing technique for laundry detergent products is relatively new but is becoming much more significant as in the powder product market moves toward concentrated products.

Agglomeration involves the contacting of dry materials with liquid binder to form granulated particles, or agglomerates. It is normally practiced on a continuous basis. The composition of each granule is then generally representative of the entire formulation. A comprehensive review of detergent agglomeration equipment and processing is provided by Dolan [55].

In a typical detergent agglomeration process, liquid sodium silicate is sprayed onto a moving bed of blended dry raw materials. Hydratable components of the bed, such as sodium tripolyphosphate and sodium carbonate, dehydrate the silicate solution and "stick" the particles together to form roughly spherical granules or agglomerates. Once formed, the granules are usually conditioned or further dried, resulting in crisp, free-flowing products with rather uniform particle sizing. Heat- or moisture-sensitive components of the detergent can be postadded via spray or in-line mixing as described for spray-dried products. Bulk densities of agglomerated products normally range between 0.5 and 1.0 g/cm^3, with most products residing in the upper end of this range. Figure 6.6 provides a general flow diagram of detergent agglomeration processing.

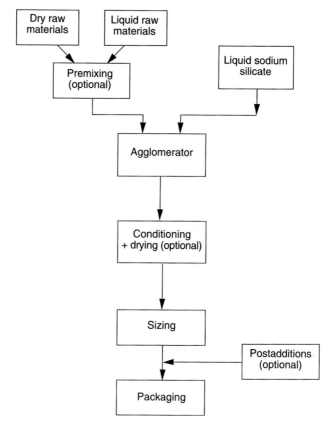

Figure 6.6   Block diagram of generic detergent agglomeration processing.

A variety of equipment has been used for the agglomeration processing of automatic dishwash compounds, laundry detergents, or both. Today the best known and most widely used device is the horizontal rotary drum agglomerator. In this equipment, the agglomeration step is conducted by spraying the liquid silicate onto a falling curtain of dry materials that is formed from the rotation of the horizontal drum. The agglomerates are further shaped as they travel through the drum, and conditioning takes place in a second rotating drum, purged with ambient or heated air. Processing time through this equipment can range from 40 to 60 minutes. Figure 6.7 is a flow diagram of this process.

Vertical agglomeration, as practiced in a Schugi unit, sharply contrasts to the horizontal method. In this equipment the dry powder falls by gravity into a mixing chamber equipped with rapidly rotating blades. The liquid silicate is sprayed into the turbulent zone to form the agglomerates. Residence time in the mixing chamber is only a few seconds. The damp agglomerates discharge by gravity into a continuous fluidized-bed dryer/conditioner, where the hydration reactions are completed and surface moisture is driven off. Figure 6.8 is the flow diagram for a typical vertical agglomeration process.

Other pieces of equipment that have been used for laundry detergent agglomeration include the pan or disk agglomerator, zigzag blender, and rotary cone agglomerator. Equipment designs and modifications of existing equipment continue to evolve as this type of processing becomes more popular. A variation of agglomeration processing, called spray mixing, is also described in the literature [2].

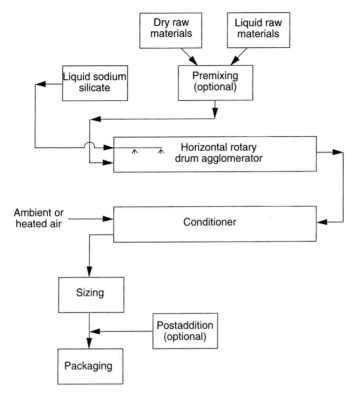

Figure 6.7   Block diagram of a horizontal rotary drum agglomeration process.

Agglomeration occupies an intermediate position between spray-drying and dry blending from a production volume perspective. It has many of the processing attributes of a spray tower but is less demanding of capital and energy. It gives better uniformity and aesthetics than dry mixing, rivaling spray-drying on these criteria. Figure 6.9 compares the appearance and uniformity of products made by these three processing techniques. Agglomeration equipment does not fully duplicate the flexibility of dry blending, but it is more amenable to multiple products than a spray tower. As noted previously, both automatic dish and laundry products can be readily made in the same agglomeration unit.

6.4.2.4   Alternative Production Techniques

The introduction of concentrated dry detergents to the detergent market has required development of alternative processing technologies. Combinations of existing processing methods are being applied while new approaches continue to be developed. Lee [56] provides a description of developing process technology for the new products.

Review of the patent literature and physical examination of the early concentrated laundry detergent products indicate multistep processing techniques. Typically a manufacturer may spray-dry a detergent-based bead, densify the spray-dried particles in a milling step, agglomerate or granulate with surfactant, and finally postadd any sensitive components, such as enzymes. The milling (densifying) and agglomeration/granulation are done in the

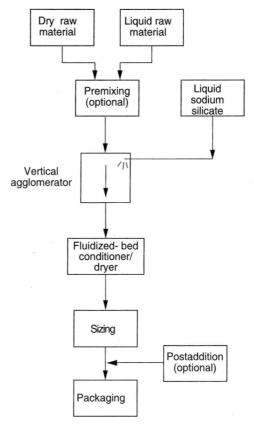

Figure 6.8   Block diagram of a vertical agglomeration process.

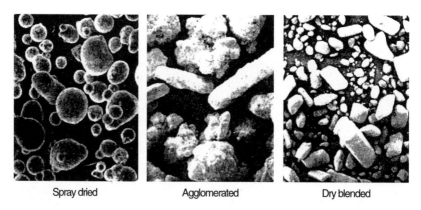

Spray dried   Agglomerated   Dry blended

Figure 6.9   Photomicrographs of detergents made by alternative processing techniques (50X).

same piece of equipment in some adaptations of this hybrid processing. Recent patent art describes a variety of innovative approaches to the manufacture of concentrated detergents without the use of a spray tower (compaction, granulation, extrusion, etc.). Photomicrographs of some typical concentrated detergent products are provided in Fig. 6.10.

Figure 6.10   Photomicrographs of typical concentrated detergent products (25X).

In general, the aesthetic properties of concentrated powders are poorer than those of conventional spray-dried or agglomerated products. They have a wide range of particle sizes, exhibit poorer flow and caking characteristics, and tend to dissolve more slowly. Some of these issues will undoubtedly be resolved as new technology for production of concentrated powders evolves.

6.4.2.5   Summary

Spray-drying, agglomeration, and dry mixing continue to serve as the principal techniques for producing powdered laundry detergent products. Hybrid and alternative processing technologies are being developed for production of concentrated detergent products. Each process makes a distinctive type of product, with its own inherent properties. Some products can be made effectively in only one type of process.

Selection of the right equipment and process design for individual producers require a careful analysis of many variables and usually involve a series of tradeoffs. The variables include type and number of products, volume requirements for each product, physical characteristics desired, formulations, raw material selection and availability, capital cost, and operating costs. Table 6.5 provides a qualitative comparison of the alternative dry detergent processing methods. In most cases the selection hinges on flexibility needs, volume requirements, desired product characteristics, and cost.

Table 6.5   Comparison of the Principal Dry Detergent Processing Techniques

Processing factors	Blending	Dry agglomeration	Spray-drying
Equipment considerations			
Operational mode	Batch	Continuous	Continuous
Size/space (relative)	Small	Intermediate	Large
Production volume	Low	Intermediate	High
Flexibility (product variety)	High	Intermediate	Low
Raw material choices	Limited	Intermediate	High
Product attributes			
Appearance/uniformity	Good	Better	Best
Bulk density	Higher	Intermediate	Lowest
Sizing (relative)	Variable	Larger	Smaller
Solubility	Slowest	Intermediate	Fastest
Cost considerations			
Capital	Low	Higher	Highest
Energy	Low	Intermediate	Highest

6.4.3   Liquid Detergent Processing

The production of clear liquid detergent products is relatively easier than the manufacture of their dry counterparts and is typically done in a batchwise mode. Some newer facilities use continuous processing for high-volume products. However, liquid processing is also becoming more sophisticated with the advent of slurries and gels for automatic dishwashers and the introduction of structured and nonaqueous products to the laundry detergent market.

6.4.3.1   Clear Liquid Products

The majority of liquid laundry products sold in North America and Europe are isotropic liquids, based on solubilizing surfactants and other formulation ingredients, with hydrotropes and solvents. These detergents require careful blending of raw materials to produce clear, stable products. Raw material quality, order of raw material addition, temperature control, and thorough mixing are important variables in obtaining consistent detergent products batch to batch. Key final product properties include clarity, viscosity, and storage stability. Most of these properties are dictated by the formulations, although improper processing can result in problems in any of these areas.

6.4.3.2   Structured and Nonaqueous Liquids

Structured liquids enjoy some measure of popularity in Europe. These products have the ability to suspend high levels of finely divided solids through formation of lamellar phase droplets of surfactant-rich material. Nonaqueous, or essentially water-free liquid detergent products, are also beginning to appear in selected world markets.

The principal difference between production of clear liquids and alternative liquid product forms lies in the mixing requirements. In general, high shear equipment is used to produce these liquids, and the order in which raw materials are added becomes a more important consideration. The processing technology for these newer liquid product forms, like that for dry concentrates, is evolving.

6.4.4   Typical Formulations

Tables 6.6–6.9 provide typical dry and liquid consumer laundry detergent formulations. The dry formulations are grouped according to manufacturing technique. In general, the dry

Table 6.6   Typical Dry Mix and Dry Neutralization Laundry Detergent Formulations

Ingredients (%)[1]	Dry mix		Dry neutralization[2]	
	P	No-P	P	No-P
Sodium tripolyphosphate	20–36	—	20–36	—
Sodium carbonate	10–30	20–50	10–30	20–50
Alcohol ethoxylate	6–10	6–10	2–4	2–4
Sodium alkylbenzenesulfonate	0–3	0–3	—	—
Dodecylbenzenesulfonic acid	—	—	8–12	8–12
Sodium silicate, hydrous	6–10	6–10	6–10	6–10
Sodium sulfate	Balance	Balance	Balance	Balance
Carboxymethylcellulose	0.5–2.0	0.5–2.0	0.5–2.0	0.5–2.0
Fluorescent whitening agent	0.1–0.5	0.1–0.5	0.1–0.5	0.1–0.5
Perfume	0.1–0.3	0.1–0.3	0.1–0.3	0.1–0.3
Water			2–4	1–3

[1] Enzymes and/or a peroxygen bleach such as sodium perborate are optional ingredients.
[2] Order of ingredient addition during processing is critical to success using this processing approach.

Table 6.7   Typical Laundry Detergent Formulations Made by Agglomeration

Ingredients (%)[1]	P type	No-P type
Sodium tripolyphosphate	20–36	—
Sodium carbonate	10–30	15–30
Zeolite A	—	15–25
Alcohol ethoxylate	6–10	6–10
Sodium alkylbenzenesulfonate	0–4	0–4
Sodium silicate (solids)	5–10	2–6
Sodium sulfate	Balance	Balance
Carboxymethylcellulose	0.5–2.0	0.5–2.0
Fluorescent whitening agent	0.1–0.5	0.1–0.5
Perfume	0–0.3	0–0.3
Water	3–6	1–3

[1] Enzymes and/or a peroxygen bleach such as sodium perborate are optional ingredients.

Table 6.8   Typical Spray-Dried Laundry Detergent Formulations

Ingredients (%)[1]	Nonionic surfactant based		Sodium LAS based		Mixed surfactant	
	P	No-P	P	No-P	P	No-P
Sodium tripolyphosphate	20–36	—	20–36	—	20–36	—
Sodium carbonate	10–30	15–30	10–30	15–30	10–30	15–30
Zeolite A	—	15–25	—	15–25	—	15–25
Sodium alkulbenzenesulfonate	—	—	10–18	12–20	5–12	5–12
Sodium alcohol ether sulfate	—	—	—	—	2–8	2–8
Alcohol ethoxylate	6–10	6–10	—	—	1–3	1–3
Sodium silicate (solids)	5–10	3–6	5–10	3–6	5–10	3–6
Sodium sulfate	Balance	Balance	Balance	Balance	Balance	Balance
Enzymes	0–2	0–2	0–2	0–2	0–2	0–2
Polycarboxylate polymer	0–2	1–4	0–2	1–4	0–2	1–4
Fluorescent whitening agent	0.1–0.5	0.1–0.5	0.1–0.5	0.1–0.5	0.1–0.5	0.1–0.5
Perfume	0–0.3	0–0.3	0–0.3	0–0.3	0–0.3	0–0.3
Water	3–8	2–4	3–8	2–4	3–8	2–4

[1] Peroxygen bleach, bleach activator, and fabric softener/antistat products are optional ingredients.

Table 6.9   Typical Liquid Laundry Detergent Formulations

Ingredients (%)[1]	Formulation	
	Built	Nonbuilt
Sodium alkylbenzenesulfonate	5–15	0–10
Alcohol ether sulfate	0–15	0–12
Alcohol ethoxylate	5–10	15–35
Sodium citrate	5–10	
Hydrotrope(s)	2–6	2–6
Solvent	2–8	2–8
Water	Balance	Balance
Enzyme	0–2	0–2
Fluorescent whitening agent	0.1–0.3	0.1–0.3
Perfume	0–0.2	0–0.2

[1] Fabric softeners/antistat agents and coloring agents are optional ingredients.

formulations exhibit increasing complexity that corresponds to the wider choice of raw materials available for use in the high-volume spray-dried products, compared to the rather simple dry mix and slightly more sophisticated formulations made via agglomeration.

The typical formulations are expected to provide good to excellent performance in the hands of the consumer. The optional ingredients add cost and functionality to the products, as well as a means to differentiate products and performance claims in the market.

6.5   Product Form and Packaging

6.5.1   Introduction

One way in which producers can differentiate their products from those of their competitors is modification of the product form offered to the consumer. Closely associated with product form are packaging alternatives, which can be complementary and, in some cases, can add new dimensions and points of difference to purchase selection. Product form, like formulations and processing, is in a state of change driven largely by growing consumer environmental awareness, changing demographics (particularly in the highly industrialized countries), changing wash conditions and habits, cost pressures, and fierce competition for market share among the major worldwide detergent producers.

Significant changes related to product form occurred in the U.S. detergent market in 1984, when liquid detergents made major share gains and again in 1991, following the introduction of concentrated detergents the preceding year. Figure 6.11 shows market share of spray-dried, liquid, and concentrated detergents between 1980 and 1992.

6.5.2   Dry Products

The largest variety of product forms and packaging options lies in the dry product category. The degree of sophistication ranges from dry mixed products packed in cardboard cartons through products offered in multicomponent pouches, with many options in between.

6.5.2.1   Spray-Dried Detergents

Spray-dried laundry detergents are the standard product form and represent the largest volume of detergent products offered around the world. Packaging of these products is

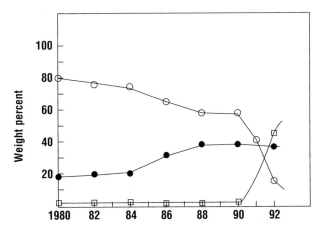

Figure 6.11   Product form distribution of U.S. laundry detergents: ○, spray-dried powders; □, concentrated powders; ●, liquid detergents.

typically in cardboard cartons or in plastic bags of various sizes. Only within the past few years has any alternative dry product form made a significant impact on market share.

Spray-dried products are characterized by their relatively low bulk density, relatively high recommended use volume, and generally good solubility characteristics. Bulk density of products around the world generally ranges between 0.3 and 0.6 g/cm^3, with the European products generally in the high end of this range. Recommended use volume is typically 0.75 to 1.5 cups of detergent per wash load. Producers have adjusted bulk density and made changes in recommended use level to differentiate their products and, in effect, to provide a means of "controlling" the amount of detergent used in the wash load.

6.5.2.2   Tablets and Bars

Detergent tablets have been introduced to the market from time to time with little success. The principal shortcomings of this product form have been product integrity (breakage during processing, packaging, and shipping), solubility problems, and lack of flexibility in adjusting dosage to the wash load.

Detergent bars of various shapes remain a very popular product form in developing countries, where washing clothes by hand is prevalent. Markets are found in Latin America, Africa, and parts of Asia, including China. Bar hardness is a key property, because the bar is typically rubbed directly on the clothing and softer bars dissolve too rapidly.

6.5.2.3   Pouches and Sheets

Detergents packaged in individual wash-load sized pouches have also been introduced to the market over time with no lasting success. These products have been offered in both water-soluble pouches and disposable fabric varieties. The soluble pouches have been filled with dry detergent or, in one case, a slurry.

The fabric pouches offered more opportunities for differentiation and typically were used to deliver detergent to the washing machine and antistat/fabric softener in the clothes dryer. Perhaps the most sophisticated pouch contained separate compartments for laundry detergent and dry bleach, and the fabric of the pouch itself was impregnated with antistat/fabric softener. This multifunctional product, which had the potential to replace three separately purchased cleaning products, did not advance beyond test marketing status in the United States. While it was speculated that production and retail cost were major issues, this product also limited the opportunity for the consumer to adjust wash load dosage and removed the option of omitting bleach from the wash load. Dryer softening remained optional, since the pouch could be discarded after the wash cycle.

At least one detergent sheet was also introduced to the U.S. detergent market. A recent product took the form of a nonwoven fabric impregnated with the cleaning ingredients. Again, consumer dosage control was an issue, as was consumer concern that adequate cleaning performance could be obtained with this product form.

6.5.2.4   Concentrated Products

Concentrated laundry detergents have made significant market share inroads in the Japanese, European, and U.S. markets in the past few years. This product form is now being offered in most world markets. Prior introductions of concentrated product met with limited success, perhaps because consumers suspected that lower volume use recommendations were insufficient to achieve adequate cleaning or because the new concentrates cost more per pound than spray-dried products.

The beginning of the current era of concentrated detergents is attributed to the Attack™ brand developed by Kao Corporation (Japan) in early 1987. This product combines cleaning performance based on novel enzyme technology with increased density (0.70 g/cm^3) and new, convenient small size packaging.

In a very short time, concentrated products became the preferred product form in Japan. European producers introduced concentrates to their consumers, and the United States was close behind. While acceptance of these products in Europe and North America has been slower than in Japan, this product form now commands significant market share in both areas.

The wide acceptance of the concentrated product form can be attributed to a number of consumer-related factors. First, the performance is judged to be at least equivalent to that of conventional products. Second, the products are viewed as being very convenient in terms of having "user-friendly" packaging (small boxes that won't tip over easily, handles, measuring scoop included). Finally, these products appeal to consumers' sense of environmental responsibility: less detergent is used per wash, less packaging material is used, the cartons are made from recycled paper, and the handle and scoop can be recycled.

Concentrates also offer significant advantages to detergent producers and retailers. The producers' benefits include a new platform for differentiation, lower raw material usage, reduced shipping costs for the same number of wash loads, lower requirements for packaging materials, and greater processing flexibility. The retailer handles fewer numbers of cartons and gains shelf space.

Rapid changes are occurring in detergent concentrates as producers compete for shelf space and market share. New formulations are being introduced, new processing technologies are being employed, and a variety of concentrates and superconcentrates are available to the consumer. As noted earlier, concentrates have been introduced to consumers throughout the world. At present, acceptance has been mixed outside Japan, Europe, and North America.

6.5.3   Liquid Products

Liquid product forms have some advantages over dry products, including ease and cost of manufacture, attractive appearance, variety of container shapes available, rapid solubility, and ease of fabric pretreatment. The disadvantages, relative to dry products, include some formulation limitations, solubility limitations of certain ingredients, and formulation sensitivity to extreme storage environments. The rapid solubility of liquids and convenience of use are frequently cited as the key factors in their significant acceptance in the North American market.

Historical liquid laundry product forms were typically clear and formulated with or without builders. Typical use recommendations for U.S. liquids are 0.25 cup for nonbuilt products and 0.5 cup for built products. The trend to concentrates in dry products is carrying over to liquids, with new, more concentrated, liquids now being test marketed in North America and in Europe.

In Europe, the liquid product form has not been as well received as in North America. This may be partially due to the common use of front-loading machines on the continent versus top-loaders in the United States and Canada. However, a wider variety of liquid product forms is available in Europe, including clear liquids, structured liquids, and, more recently, concentrated nonaqueous liquids containing bleach.

6.6   Performance Testing

6.6.1   Introduction

Detergent performance testing covers a wide range of options, ranging from relatively simple laboratory methods using artificial soils under closely controlled conditions, to placement of the detergent products in consumer homes. Typically, performance testing involves a series of tests, since no single test can completely define all aspects of detergent performance. Cost of the evaluations increases, and the testing becomes more complex and subject to more variation, in moving through the various stages leading to the ultimate consumer test. Several discussions of laundry detergent testing are found in the literature [2,49,58].

6.6.2   Laboratory Testing

Laboratory testing, which can be considered to be the screening level or direction-finding level of performance testing, is usually performed with artificially soiled or stained fabrics under carefully controlled conditions (detergent concentration, wash temperature, water hardness, etc.). Terg-O-Tometers, miniature agitator washers (U.S. Testing Co.), are extensively used for laboratory benchtop testing in the United States. Reflectance measurements taken on soiled fabric swatches before the wash and again on the same swatches after the wash are used to calculate soil removal (SR) or detergency performance as follows:

$$\% \text{ SR} = \frac{R_f - R_s}{R_0 - R_s} \times 100 \tag{6.1}$$

where R_0 is the reflectance of the unsoiled fabric, R_s is the reflectance of the soiled fabric, and R_f is the final reflectance of the fabric after the wash.

It is not uncommon for laboratory tests to yield conflicting results, since artificial soils can respond differently to various formulation components. For example, soils that respond well to surfactants may not respond in a similar fashion to enzyme or bleach ingredients, and vice versa. Thus some caution is necessary in extrapolating from the results of one type of laboratory testing to arrive at predictions of the performance a consumer might obtain.

Single-cycle wash tests are routinely conducted to measure primary detergent effects, such as soil and stain removal. Multiple-cycle testing is used to gain more insight into secondary, but still very important, performance characteristics such as soil redeposition, dye fading, encrustation, and fiber damage. Special test protocols (presoaking, pretreating, etc.) can also be readily conducted at this scale. An excellent compendium and bibliography of laboratory test methods for evaluating the performance of various detergent and cleaning products, including laundry detergents, is available through the Chemical Specialties Manufacturers Association [59].

6.6.3   Practical Testing

Practical testing is typically conducted in full-sized home laundering equipment with naturally soiled laundry items. This is sometimes referred to as bundle testing. The washing and performance evaluation are done in the laboratory by trained personnel. Although reflectance measurements can be used in conjunction with this method, more typically a trained panel is used to evaluate the cleaning results. As in the benchtop testing, wash conditions can be closely controlled in this laboratory setting. The key advantages to practical testing are the use of naturally soiled laundry items and multicycle testing.

One methodology for conducting this test is detailed by the American Society for Testing and Materials in ASTM D 2960-89, Standard Test Method of Controlled Laboratory Test Using Naturally Soiled Fabrics and Household Appliances [60].

6.6.4   Consumer Testing

The ultimate test of a new detergent product, or change in the formulation of an existing product, is to put it in the hands of the consumer [61]. This type of testing is normally done on a blind basis, and the consumer is asked to compare the performance of two detergent products. Consumer testing can be used to gain insight into a wide variety of performance variables beyond simply cleaning performance, such as perfume selection, product form, color, solubility, and packaging. Needless to say, this type of testing is the most subjective, in that use conditions are subject to the laundry habits of the individual consumer. Additionally, evaluation of a detergent's overall performance is subject to the consumer's individual preferences. For example, if a product cleans well but does not have a pleasing odor, a consumer may opt for a different brand. In fact, some consumers prefer products with no perfume at all.

6.7   Detergent Usage and Cost Performance

Detergent packages normally provide recommendations on the amount of product to be used for optimum performance. These recommendations are typically based more on product form (density) than on the formulation, in that formulation adjustments are usually made within the context of a specific product form. The interesting paradox in connection with detergent usage is that these products are sold on a weight basis but used on a volume basis. This makes it difficult for the average consumer to accurately calculate and compare the wash load cost among detergents. Cost per wash calculations are further complicated if the consumer does not accurately measure the amount of detergent used or, to obtain the cleaning performance desired, ends up underusing or overusing the product.

Comparing the cost performance of powders with liquids is even more difficult, since liquid detergents are sold by the fluid ounce.

When most powder detergents were spray-dried and recommended use levels were similar, the consumer could compare prices of similar sized cartons. The introduction of concentrates removed the ability of the consumer to compare powders on a weight basis, however, because the recommended use levels differ. In an effort to mitigate the cost-per-

Table 6.10   Typical U.S. Laundry Detergents: Use Considerations

Product type	Bulk density (g/cm^3)	Recommended use volume		Detergent use concentration	
		Cups or scoop[1]	cm^3	g/wash	wt %[2]
Powders					
Dry mix	0.9	$\frac{1}{2}$ cup	120	108.0	0.168
Agglomerated	0.74	$\frac{3}{4}$ cup	180	133.2	0.207
Spray-dried	0.36	1 cup	240	86.4	0.134
Concentrate	0.55	1 scoop	120	66.0	0.103
Superconcentrate	0.95	1 scoop	80	76.0	0.118
Liquids					
Nonbuilt	1.03	$\frac{1}{4}$ cup	60	61.8	0.096
Built	1.10	$\frac{1}{2}$ cup	120	132.0	0.205

[1] Scoops included in concentrate detergent packages are variable in size depending on product density.
[2] Assumption: 17-gallon washing machine fill volume.

pound differences between spray-dried products and concentrates, the producers listed the number of washes at the use levels recommended on the cartons. For the first time, consumers could calculate cost per wash directly in making purchasing decisions.

Table 6.10 summarizes detergent usage information for the various laundry detergent product forms on the market. This information permits the calculation of cost per wash using retail price figures. It also allows a formulator to calculate the impact of density or formulation changes on ingredients delivered to the washing machine with various formulations and to make other useful calculations and comparisons.

References

1. Strandberg, K.W. *Soap Cosmet. Chem. Spec.*, 30–32 (Jan. 1993).
2. Jakobi, G., Lohr, A. *Detergents and Textile Washing,* VCH, Weinheim, Germany, 1987.
3. Perenich, T.A., Moore, M.A. *Soap Cosmet. Chem. Spec.,* 40–52 (Aug. 1988).
4. Webb, J.J., Obendorf, S.K. Text. Res. J., *57* 640–646 (1987).
5. Milwidsky, B. HAPPI, *26* 77–82 (May 1989).
6. Greek, B.F. Chem. Engr. News, 37–60 (Jan. 29, 1990).
7. Smulders, E.J. HAPPI, *23* 82–88 (Nov. 1986).
8. Hunter, D., Mullin, R., Kemezis, P., Roberts, M., Morris, G. Chem. Week, 5–8 (Jan. 27, 1993).
9. Mullin, R., Wood, A., Morris, G., Chynoweth, E., Roberts, M., Chem. Week, 24–44 (Jan. 29, 1992).
10. Ainsworth, S.J., Chem. Engr. News, 27–63 (Jan. 20, 1992).
11. Cahn, A., J. Am. Oil Chem. Soc., *65* 1367–1373 (1988).
12. Carson, H.C., HAPPI, *27* 41–48 (June 1990).
13. Milwidsky, B., HAPPI, *27* 50–52 (Feb. 1990).
14. Donohue, J., Soap Cosmet. Chem. Spec., 28–58 (Sept. 1988).
15. Carson, H.C., HAPPI, *24* 27–37 (June 1987).
16. Richter, H.J., Knaut, J., Performance Chem., 14–20 (July/Aug. 1989).
17. Jakobi, G., Stache, H., *Tensid–Taschenbuch,* 2nd ed. Hanser Verlag, Munich, 1981, 253–337.
18. Rock, S.L., Muck, D.L., HAPPI, *22* 56–80 (Apr. 1985).

19. Milwidsky, B. HAPPI, 52–84 (Aug. 1987).
20. Donohue, J., Soap Cosmet. Chem. Spec., 28–86 (Mar. 1984).
21. Hart, J.R., Soap Cosmet. Chem. Spec., 38–48 (May 1986).
22. Morgenthaler, W.W., *Proceedings of the Second World Conference on Detergents,* Montreux, Switzerland, October 1986.
23. Kissa, E., Dohner, J.M., Gibson, W.R., Strickman, D., J. Am. Oil Chem. Soc., *68* 532-538 (1991).
24. Parker, J. J. Am. Oil Chem. Soc., *60* 1102–1105 (1983).
25. Bolsman, T.A., Kok, B.M., Vreugdenhil, R., J. Am. Oil Chem. Soc., *65* 1211–1218 (1988).
26. Burg, A.W., Rohovsky, M.W., Kensler, C.J., CRC Crit. Rev. Env. Control, 91–120 (Apr. 1977).
27. Zussman, H.W. In *Encyclopedia of Polymer Science and Technology,* Vol. 2, 606–613.
28. Anliker, R., Hefti, H., Kasperl, H., J. Am. Oil Chem. Soc., *49* 75–80 (1969).
29. Stensby, P.S., Findley, W.R., Soap Cosmet. Chem. Spec., 54–64 (Oct. 1972).
30. Stensby, P.S., Soap Cosmet. Chem. Spec., 41–103 (Apr. 1967).
31. *Kirk-Othmer Encyclopedia of Chemical Technology,* Vol. 9, 3rd ed. Wiley, New York, 1980, 138–148.
32. Starace, C.A., Soap Cosmet. Chem. Spec. (May 1983).
33. Fujii, T., Tatara, T., Minagawa, M., J. Am. Oil Chem. Soc., *63* 796–799 (1986).
34. Kravetz, L., Guin, K.F., J. Am. Oil Chem. Soc., *62* 943–948 (1985).
35. Crossin, M.C., Soap Cosmet. Chem. Spec., 52–94 (Aug. 1987).
36. Christensen, P.N., Holm, P., Sonder, B., J. Am. Oil Chem. Soc, *55* 109–113 (1978).
37. Nielsen, M.H., Jepsen, S.J., Outtrup, H., J. Am. Oil Chem. Soc., *58* 644–649 (1981).
38. Tatara, T., Fujii, T., Kawase, T., Minagawa, M., J. Am. Oil Chem. Soc., *62* 1053–1061 (1985).
39. Maase, F.W.J.L., van Tilburg, R., J. Am. Oil Chem. Soc., *60* 1672–1675 (1983).
40. Gormsen, E., Malmos, H., HAPPI, *28* 122–125 (Oct. 1991).
41. Leikhim, J.W. *Proceedings of the Second World Conference on Detergents,* Montreux, Switzerland, Oct. 1986, 189–193.
42. Stinson, S.C. Consumer preferences spur innovation in detergents, Chem. Engr. News, 21 (Jan. 26, 1987).
43. Brenner, T.E. *Proceedings of the Second World Conference on Detergents,* Montreux, Switzerland, Oct. 1986, 22–27.
44. Davidson, A.S., Milwidsky, B. *Synthetic Detergents,* 7th ed. Wiley, New York, 1987.
45. Sosis, P., Burch, W.D. Role of oily soils in detergency tests, Soap Cosmet. Chem. Spec., 32 (July 1973).
46. Arai, H., Maruta, I., Kariyone, T. Study of detergency, effect of sodium tripolyphosphate, J. Am. Oil Chem. Soc., *43* 315 (1965).
47. Spangler, W.G. et al. A laboratory method for testing laundry products for detergency, J. Am. Oil Chem. Soc., *42* 723–727 (1965).
48. American Society for Testing and Materials. *1983 Annual Book of ASTM Standards, Evaluating Stain Removal Performance in Home Laundering,* Method D 4265-83, ASTM, Philadelphia, 1983.
49. Cutler, W.G., Davis, R.C., Eds. *Detergency: Theory and Test Methods.* Dekker, New York, 1972.
50. Schick, M.J., Ed. *Surface Characteristics of Fibers and Textiles.* Dekker, New York, 1977.
51. Donohue, J. Evaluation of cleaners, laundry detergents, Soap Cosmet. Chem. Spec., 50 (Feb. 1987).
52. Kent, J.A., Ed. *Riegel's Handbook of Industrial Chemistry,* 8th ed. Van Nostrand Reinhold, New York, 1983.
53. Masters, K. *Spray Drying,* 2nd ed. Halsted Press (Wiley), New York, 1976.
54. Pietsch, W. *Size Enlargement by Agglomeration.* Wiley, Chichester, 1991.
55. Dolan, M.J., HAPPI, *24*(4) 64 (1987); Soap Cosmet. Chem. Spec., *63*(3) 45 (1987).
56. Lee, A.E., HAPPI, *29*(11) 92 (1992).
57. Davidson, A., J. Am. Oil Chem. Soc., *55*(1) 134 (1978).
58. Cahn, A., Lynn, J.L. Surfactants and detersive systems, in *Kirk-Othmer Encyclopedia of Chemical Technology,* Vol. 22, 3rd ed. Wiley, New York, 1983, 387.
59. *Detergents Division Test Methods Compendium,* 2nd ed. Chemical Specialties Manufacturers Association, Washington, DC, 1985.
60. American Society for Testing and Materials. *1991 Annual Book of ASTM Standards,* Vol. 15.04, Method D 2960-89, ASTM, Philadelphia, 1992.
61. Mills, K.L., Gladstone, G., Neiditch, O.W., Chem. Times Trends, *10*(2) 50 (1987).

CHAPTER 7

Hard-Surface Cleaners

Timothy C. Morris

Hard-surface cleaning is a broad subject encompassing a multitude of cleaning applications. This chapter addresses hard-surface cleaning by discussing the following application areas involved in nonporous and, in some cases, semiporous hard surfaces:

In situ cleaners
Dishwash detergents
Vehicle wash cleaners
Metal cleaners

   These hard-surface cleaning applications are presented in four sections that include descriptions of the applications as well as the surface and soil types encountered. The function of typical cleaner ingredients is explained, based on the cleaning requirements of the application, and formulating strategies to overcome product instabilities are presented. Each section and subsection ends with examples of typical cleaner and detergent compositions to illustrate the systems discussed and to serve as starting-point formulas.

7.1   Introduction

This chapter addresses hard-surface cleaner systems designed specifically for use in both households and commercial operations. Consumer hard-surface cleaners are usually lower in active ingredients and alkalinity than their industrial and institutional (I&I) counterparts. Consumer products rely on the mechanical action of manual scrubbing in combination with the chemical cleaning activity of surfactants (surface active agents), organic solvents, chelating agents and, to some extent, alkalies. Industrial and institutional hard-surface cleaners are similar to the consumer products; however, they are usually moderately to highly alkaline, using higher levels of solvents and nonionic surfactants to remove heavier loads of grease and other hydrophobic soils. Some I&I hard-surface cleaning applications also use machines (e.g., floor scrubbers) in place of manual scrubbing to provide increased physical energy for better soil removal.

   In general, hard-surface cleaners are formulated to remove a wide variety of soils, including food (carbohydrates, fatty acids, proteins), petroleum greases and oils, body fluids, and clay and carbon particulates from many different hard surfaces. Examples of these surfaces are glass, ceramic, vinyl and acrylic, marble, metal, wood, and cement. These surfaces are found almost everywhere inside and outside homes and businesses, as mirrors

Timothy C. Morris, PQ Corporation, Research & Development Department, Conshohocken, Pennsylvania 19428, U.S.A.

and windows, sinks and bathtubs, bathroom and kitchen fixtures, kitchen appliances and oven interiors, table and counter tops, floor coverings, patios, and garage and basement floors [1].

When formulating a hard-surface cleaner, the main objective is not only to remove soil but also to leave the surface intact and, if possible, free from cleaner residue. If the surface is not rinsed, the cleaner remaining there should not leave stains, streaks, or marks. To achieve good cleaning without surface degradation and with minimal residue, cleaners need to be formulated for particular surfaces and soils. For example, spray-on kitchen cleaners are formulated to remove food stains with effective but easy-to-remove surfactants and solvents at low concentrations. Such cleaners usually are formulated with alkaline ingredients (i.e., hydroxides, carbonates, and phosphates) at very low levels, to avoid corrosive chemical interaction with glass, ceramic, and metal surfaces typically encountered. Undesirable chemical corrosion is often exhibited as a hazing of the surface, caused by the dissolving or "etching" effect of the alkaline solution on the substrate being cleaned.

Other important formulating considerations are the product's impact on the health and safety of the end user and on the environment. For instance, the use of surfactants that fully biodegrade to harmless components is preferred. Also, solvents that are classified as nontoxic should be used. For consumer products, high-pH cleaner systems, which are corrosive to human skin and eyes, should be limited to difficult applications, such as oven cleaning.

In recent years, hard-surface cleaners have come mainly in liquid form; however, powdered hard-surface cleaners are still popular. Liquid hard-surface cleaners are marketed as either low viscosity, low concentration liquids applied as spray (aerosol or pump), or as concentrated liquids having a relatively high viscosity. The higher viscosity cleaners usually are diluted before use. The powdered products, mainly sold as heavy-duty, general-purpose cleaners, are concentrated. These contain alkaline salts, such as sodium phosphates, carbonates, and silicates, as well as surfactants and solvents. Unlike the liquid products, powdered hard-surface cleaners must be dissolved before use.

7.2   In Situ Cleaners (Clean-in-Place)

7.2.1   Conventional Hard Surfaces

In situ cleaners are typically formulated as ready-to-spray liquids (pump or aerosol), or as liquid concentrates that are diluted before use. They are designed primarily to remove petroleum and fatty acid containing greases and oils, clay and other particulate soils, and hard water salts, such as soap scum. Applications include bathroom tubs, sinks, and tiles and kitchen floors and appliances. These cleaners rely on surfactants, solvents, and alkalinity to cut through grease and help remove dirt. Sequestering agents are also incorporated to remove water hardness ions (i.e., calcium and magnesium), enabling surfactants to perform optimally. To some degree, they also solubilize hard water salts, such as calcium carbonates and calcium/magnesium soaps (soap scum).

Anionic surfactants are the predominant surface active agent used in these formulations because of their excellent cost/performance on oily and particulate soils. Also, anionic surfactants generally exhibit better product stability and easier removal from surfaces compared with surfactants of other types (i.e., nonionic). The disadvantage of anionic surfactants is their strong interaction with the calcium and magnesium cations naturally

Table 7.1   Relation of HLB Number to Surfactant Function

HLB range	Use
4–6	Water-in-oil emulsifiers
7–9	Wetting agents
8–18	Oil-in-water emulsifiers
13–15	Detergents
10–18	Solubilizers

Source: *The Atlas HLB System,* 4th printing, Atlas Chemical Industries (now Imperial Chemical Industries), Wilmington, DE, 1963.

present in water. This interaction reduces the anionic surfactant's solubility, rendering it less effective in removing soils. One commonly used anionic surfactant is linear alkyl-benzenesulfonic acid (LAS) neutralized with sodium hydroxide, ammonium hydroxide, or an alkylolamine, or with a combination of these bases [1]. Other anionic surfactants commonly used include the alkyl sulfates (AS) and the alkyl ether sulfates. Alkyl ether sulfates and, to some degree, AS are more effective in removing soils than LAS in hard water (high levels of Ca^{2+} and Mg^{2+} ions) and tend to be milder to skin.

Nonionic surfactants, such as the ethoxylated linear alcohols [2] and the ethoxylated alkylphenols, are also incorporated to remove oily soils, especially petroleum-based, and to help reduce the level of foam generated by anionic surfactants. Although polar in nature, nonionic surface active agents do not ionize in solution and are, therefore, less electrostatically interactive with metal cations such as Ca^{2+} and Mg^{2+}. Nonionic surfactants are generally less soluble in water than anionic surfactants and, consequently, more difficult to stabilize in a liquid formulation. This is especially true if the cleaner is highly concentrated. Often a hydrotrope, which is a chemical that increases the aqueous solubility of a slightly soluble organic compound, is used. Common hydrotropes used in liquid in situ cleaners include sodium xylene, toluene or cumene sulfonates, and organic phosphate esters. These organic compounds are anionic surface active agents that will solubilize not only other surfactants but slightly water-soluble solvents as well.

When describing a surfactant function, whether as an emulsifier, a wetting agent, or a detergent, the HLB system is often used. HLB stands for hydrophile (water-loving)/lipophile (oil-loving) balance of a surfactant. The lower the HLB number, the greater the affinity for oil. The higher the HLB number, the greater the affinity for water. Therefore, HLB numbers can be used to indicate which function a surfactant performs best. Generally, Table 7.1 relates the HLB number to a surfactant's function.

Of course, many in situ cleaner formulations use combinations of both anionic and nonionic surfactants. Laboratory testing of cleaning systems has given reasonably good evidence that a 80:20 mixture of LAS and a nonionic surfactant with an HLB of 12.5–13.2 shows promise of giving optimal results [3].

Solvents are incorporated to help remove fatty food soils which have not been saponified by the alkali present. These solvents, where possible, should be nontoxic and preferably odorless; they should have a fairly high flash point and good fat solvency. Glycol ethers, such as ethylene glycol monobutyl ether, are the most widely used in situ cleaner systems. Because the toxicity level of ethylene oxide based glycol ethers has been questioned, the safer propylene analogs may be more appropriate [4]. Another solvent used is isopropyl alcohol

(IPA) or isopropanol. IPA is used as a hydrotrope as well as for fat solvency [5]. Other low molecular weight alchohols used as sovents and hydrotropes are methanol and ethanol. When formulating with these volatile solvents, care should be taken to produce a finished product with a safe flash point. Less water-miscible solvents frequently include pine oil and the relatively new D-limonene. Pine oil, extracted from wood, is very good on oily food soils of many types and also possesses bactericidal effects [5]. D-Limonene, which is a naturally derived terpene extracted mainly from citrus, has excellent solvency and will impart a pleasant citrus odor to the cleaner.

Alkalinity is an important property of most formulations. The ability of a cleaner to neutralize fatty soils and, in some cases, saponify them for better removal, will depend on the presence and concentration of alkaline ingredients. Common sources of alkalinity include sodium hydroxide, sodium carbonate, sodium orthophosphate, and their potassium analogs. Nitrogen-based alkalies, such as ammonia and the various alkanolamines (e.g., mono- and triethanolamine), are also used frequently. In addition, some sequestering agents, such as tetrapotassium pyrophosphate (TKPP) and nitriloacetate (NTA) trisodium salt, provide alkalinity.

Corrosion protection is important. Sodium and potassium silicates, which are also used as alkaline buffers, provide effective protection of metal, ceramic, and glass surfaces from adverse chemical interaction during cleaning.

Sequestering agents, sometimes referred to as detergent builders, are also incorporated because they "build up" the cleaner's detergency by upgrading or protecting the surfactants. Sequestering agents are compounds that form soluble complexes with multivalent metal cations. Common examples are tetrapotassium pyrophosphate (TKPP) and sodium tri-polyphosphate (STPP). These phosphate compounds are formulated into products to remove calcium and magnesium ions from the cleaning solution. Other sequestering agents commonly used are ethylenediamine tetraacetic acid (EDTA), sodium gluconate, sodium citrate, and NTA. These organic sequestering agents are effective at removing Ca^{2+} and Mg^{2+} ions and other metal ions (e.g., Fe^{2+} and Fe^{3+}) that can be detrimental to good cleaning.

When choosing a sequestering agent for water hardness removal, refer to the stability constant of the metal complexes formed. The higher the stability constant, the greater the affinity of the metal cations for that sequestering agent [6]. If a sequestering agent with a high stability constant is chosen, chances are that ingredient will also be effective at conditioning the water for better detergency. However, other considerations, such as the pH of the cleaning solution, ion removal capacity, and cost, will affect the final choice of detergent builder system. Formulations for some hard-surface (in situ) consumer and institutional applications are given below (compositions indicated in weight %):

Hard-Surface Spray Cleaner, *Description: Consumer alkaline liquid for kitchens and bathrooms. Application: Spray and wipe. Use level: As is.*

Water	90.00
Tetrapotassium pyrophosphate	2.00
Liquid sodium silicate[1]	1.88
Sodium hydroxide (50%)	0.12
Dipropylene glycol monomethyl ether	5.00
Octylphenol, 12–13 moles ethylene oxide	1.00

[1] 3.22 weight ratio, 38% solids, clear grade.

Hard-Surface Spray Cleaner Concentrate, *Description: Consumer nonphosphated liquid for kitchens and bathrooms. Application: Spray and wipe or mop. Use level: As is, or dilute 1 part cleaner with 3 parts water.*

Water	85.5
Sodium alkyl benzenesulfonate (100%)	3.8
Alcohol ether sulfate (100%)	1.2
Monoethanolamine	2.5
Ethylene glycol *n*-butyl ether	6.0
Ethylenediaminetetraacetic acid	1.0

Source: Ref. [1].

Hard-Surface Spray Cleaner Concentrate, *Description: Consumer nonphosphated aerosol spray for kitchens and bathrooms. Application: Spray and wipe. Use level: As is. Container: Lacquered tinplate with mechanical breakup value.*

Water	67.05
Propylene glycol methyl ether	13.50
Tall oil, ethoxylated (HLB 13.8)	5.40
Octyphenol, 9 moles ethylene oxide	3.60
Ammonia (0.880)	0.45
Propellant 12/114 (50:50)	10.00

Source: Ref. (4), p. 326.

Hard-Surface Spray Cleaner, *Description: Institutional alkaline liquid. Application: Spray and wipe. Use level: As is.*

Water	88.6
Sodium metasilicate, anhydrous	2.4
Sodium tripolyphosphate	2.0
Ethylene glycol *n*-butyl ether	3.0
Ethylenediaminetetraacetic acid	1.0
C_9–C_{11} linear alcohol, 6 moles ethylene oxide	3.0

Source: PQ Detergent Formulary.

Hard-Surface Cleaner Concentrate, *Description: Institutional alkaline liquid. Application: Wipe or mop. Use level: 2–8 oz./gal.*

Water	65.5
Sodium alkylbenzenesulfonate (100%)	4.0
Phosphoric acid (85%)	4.5
Potassium hydroxide (100%)	9.0
Tetrapotassium pyrophosphate	4.5
Liquid sodium silicate[1]	4.5
Ethylene glycol *n*-butyl ether	6.0
Isopropyl alcohol	2.0

[1] 2.0 weight ratio, 44% solids, clear grade.
Source: Ref. [5], p. 227.

Hard-Surface Cleaner Concentrate, *Description: Institutional alkaline nonphosphated liquid. Application: Wipe or mop. Use level: 2–8 oz./gal.*

Water	50.3
Ethylenediaminetetraacetic acid, sodium salt	7.4
Liquid sodium silicate[1]	15.0
Sodium hydroxide (50%)	7.5
Sodium xylene sulfonate (100%)	12.8
C_{12}–C_{13} linear alcohol, 6.5 moles ethylene oxide	7.0

[1] 3.22 weight ratio, 38.4% solids, clear grade.
Source: PQ Detergent Formulary.

Hard-Surface Spray Cleaner, *Description: Institutional alkaline aerosol spray. Application: Spray and wipe. Use level: As is. Container: Plain tinplate can; best pressurized with 7% butane.*

Water	84.0
Sodium metasilicate, anhydrous	2.0
Sodium tripolyphosphate	2.0
Tetrasodium pyrophosphate	2.0
Sodium alkylarylsulfonate (60%)	2.0
Sodium xylene sulfonate	2.0
Ethylene glycol *n*-butyl ether	6.0

Source: Ref. [8], p. 327.

Hard-Surface Cleaner Concentrate, *Description: Consumer alkaline liquid. Application: Wipe, mop, or spray. Use level: As is or diluted 2–4 oz./gal.*

Water	78.1
Sodium metasilicate, anhydrous	2.4
Tetrapotassium pyrophosphate (605)	3.0
Phosphate ester	3.0
Propylene glycol monomethyl ether	6.0
Octylphenol, 9 moles ethylene oxide	5.0
Octylphenol, 5 moles ethylene oxide	2.5

Source: PQ Detergent Formulary.

Hard-Surface Cleaner Concentrate, *Description: Consumer alkaline nonphosphated liquid. Application: Wipe, mop, or spray. Use level: As is or diluted 2–4 oz./gal.*

Water	62.2
Sodium alkylbenzenesulfonate (100%)	15.0
Ethylenediaminetetraacetic acid, sodium salt	4.0
Liquid sodium silicate[1]	6.8
Ammonia (100%)	5.0
Ethylene glycol *n*-butyl ether	7.0

[1] 2.00 weight ratio, 44% solids, clear grade.
Source: Ref. [5], p. 226.

Hard-Surface Cleaner Concentrate, *Description: Consumer, pine oil based, liquid. Application: Wipe or mop. Use level: 2–4 oz./gal.*

Water	59.6
Sodium alkylbenzenesulfonate (100%)	4.7
Pine oil	20.0
Isopropyl alcohol	11.0
Triethanolamine	4.7

Source: Shell Formulary.

7.2.2   Glass Cleaners

Glass cleaners are the lowest strength of all hard-surface cleaners. The goal in formulating a glass cleaner is to minimize streaks and chemical interaction with glass surfaces. Therefore, typical formulations have very low concentrations of solvents, such as isopropyl alcohol, and anionic surfactants. Often, ammonia is incorporated at low levels as a mild source of alkalinity for fatty grease removal. These cleaners are usually not very effective on kitchen appliances and other surfaces having soil loads that are heavier and more difficult to remove. Some typical glass cleaner compositions are given below (compositions indicated in weight %):

Window Cleaner, *Description: Consumer liquid. Application: Spray. Use level: As is.*

Deionized water	91.9
C12–C13 linear alcohol, 9 moles ethylene oxide	0.1
Isopropyl alcohol	5.0
Ethylene glycol *n*-butyl ether	3.0

Source: Ref. [1].

Window Cleaner, *Description: Consumer and institutional liquid. Application: Spray. Use level: As is.*

Deionized water	87.8
Sodium alkyl sulfate (30%)	2.0
Isopropyl alcohol	7.0
Ammonium hydroxide (28%)	0.2
Ethylene glycol *n*-butyl ether	3.0

Source: Stepan Product Bulletin.

Window Cleaner, *Description: Consumer and institutional aerosol spray. Application: Spray and wipe. Use level: As is.*

Deionized water	86.98
Isopropyl alcohol	4.75
Ethylene glycol *n*-butyl ether	2.85
Nonionic surfactant[1]	0.24
Ammonium hydroxide (28%)	0.18
Isobutane propellant	5.00

[1] Polyalkylene oxide modified dimethylpolysiloxane.
Source: Union Carbide Technical Report, Glycol Ethers & Formulations.

7.2.3   Oven Cleaners

Oven cleaners are formulated to remove baked-on grease, as well as other food soils, such as burnt carbohydrates and proteins. Baked-on grease is highly polymerized and difficult to remove. Therefore, to be effective, oven cleaners require a strong, active cleaning system. To meet this requirement, most oven cleaner formulations are based on one of, or a combination of, the following cleaning systems:

- High alkalinity, usually from an alkali metal hydroxide
- Concentrated solvent, such as glycol ether
- Abrasive, such as ground pumice

High alkalinity is important in removing grease via neutralization and saponification. Solvents are effective at solubilizing food soils when used with surfactants and alkaline salts. Abrasives, such as pumice powder, help to physically remove tenacious soils. Many oven cleaning systems are formulated with thickening agents for use in aerosol applications or as pastes that are painted on. Controlling the rheology of these cleaners with organic and/or inorganic thickening agents allows them to be applied without running [7]. Below are examples of all three of the above-mentiond oven cleaning systems (compositions indicated in weight %):

Oven Cleaner, *Description: Consumer low alkaline liquid with abrasive. Mixing instructions: Stir bentonite to a gel in the water. Add the glycerin, petroleum sulfonate, diethylene glycol oleate, and sodium silicate to the bentonite gel and warm. Stir in the soap chips and allow to swell and dissolve. Add the sesquicarbonate and pumice powder and stir intermittently. When the mass is homogeneous, stir in the ethylene dichloride and cool.*

Soap chips (high titer 85% fatty acids)	3.75
Crude glycerin, 80%	1.87
Petroleum sulfonate	3.00
Diethylene glycol monooleate	1.87
Sodium silicate	1.50
Sodium sesquicarbonate	1.50
Pumice powder	41.20
Bentonite	1.86
Ethylene dichloride	2.25
Water	41.20

Source: Ref. [8], Vol. 2.

Oven Cleaner, *Description: Consumer aerosol spray. Mixing instructions: Add chemicals slowly, in order listed, while mixing. Package in aerosol dispensers (standard tin-lined can, low tin solder) with neoprene gasket, 70-durometer valve.*

Water	46.0
Magnesium aluminum silicate[1]	1.5
Tripropylene glycol methyl ether	20.0
Nonylphenol, 9 moles ethylene oxide	1.0
Sodium hydroxide (30%)	12.0
Anionic surfactant (45%)[2]	15.0
Isobutane	4.5

[1] Complex colloidal thickener.
[2] Sodium dodecyl diphenyloxide disulfonate.
Source: Dow Chemical USA, *The Glycol Ethers Handbook.*

Oven Cleaner, *Description: Consumer and institutional high alkaline liquid spray. Packaging: Pump spray.*

Water	80.4
Magnesium aluminum silicate[1]	1.2
Xanthan gum	0.4
Sodium hydroxide (50%)	8.0
Amphoteric surfactant[2]	10.0

[1] Complex colloidal thickener.
[2] Sodium salt of 2-caprylic-1(ethyl-b-oxipropanoic acid)imidazoline, 50% solids.
Source: T. Vanderbilt, Technical Bulletin, VANGEL and VEEGUM: Suggested Formulation No. 324.

Oven Cleaner, *Description: Consumer and institutional high alkaline liquid spray.*

Water	69.0
High MW acrylic polymer thickener[1]	6.0
Amphoteric surfactant[2]	3.0
Potassium hydroxide (45%)	22.0

[1] Alkali-soluble high MW acrylic polymer emulsion (30% solids).
[2] Dicarboxylic coconut derivative, sodium salt (70%).
Source: Rhône-Poulenc (formerly Miranol Chemical), Technical and Product Development Data: Suggested Formulation.

7.2.4   General-Purpose Cleaners

General-purpose cleaners are a relatively broad category of detergent products used for many cleaning chores. Most of these tasks involve hard-surface cleaning, such as kitchen and bathroom floors, basements and driveways, and patios.

General-purpose cleaners are often formulated as powders, which are diluted with water before use. Some products, especially those formulated for industrial use, are considered to be heavy-duty detergents designed to handle large soil loads. Typical formulations include anionic and nonionic surfactants for wetting and soil emulsification, and sodium phosphates and carbonates for water softening and buffering. These products typically contain sodium silicates for corrosion protection, buffering, and soil dispersion. Sometimes solvents, such as pine oil or a glycol ether, are incorporated for degreasing. When high alkalinity is required for industrial-strength cleaning, sodium hydroxide and/or sodium metasilicate is used. Liquid general-purpose cleaners are also available, usually incorporating higher concentrations of surfactants and solvents, and lower levels of alkaline builders, than the powdered products. In general, the liquids are formulated to be used "as is" or diluted.

The powdered general-purpose cleaners are generally easier to formulate as highly built detergents than their liquid counterparts. This advantage is attributed to the absence of formulation incompatibility problems. Many times, solution instabilities are encountered when one is trying to combine alkaline salts with surfactants and solvents in a liquid product. These instabilities can be exhibited as crystal or precipitate formations of the alkaline salts and/or as separation of the organic components from the aqueous phase. Frequently, to avoid these problems, a liquid cleaner's critical ingredients (i.e., phosphates, carbonates, silicates) must be reduced or even eliminated. The resulting product may be less effective in hard water and on acidic soils. Below are some typical general-purpose cleaner formulations (compositions indicated in weight %):

General–Purpose Cleaner, *Description: Consumer low alkaline powder. Use level: 1–4 oz./gal.*

Sodium alkylarylsulfonate (powder)	20.0
Sodium tripolyphosphate	30.0
Sodium hydrous polysilicate[1]	15.0
Carboxymethylcellulose (100%)	0.5
Sodium sulfate	Balance

[1] 2.0 weight ratio powder, 82.5% solids.
Source: PQ Corporation.

General-Purpose Cleaner, *Description: Consumer low alkaline, nonphosphated powders. Use level: 1–4 oz./gal.*

	A	B
C_{12}–C_{15} linear alcohol, 9 moles ethylene oxide	15.0	12.0
Sodium carbonate (light density)	60.0	30.0
Sodium citrate		20.0
Sodium hydrous polysilicate (fine powder)[1]	10.0	10.0
Carboxymethylcellulose (100%)	0.5	0.5
Sodium sulfate	Balance	Balance

[1] 2.0 weight ratio powder, 82.5% solids.
Source: Shell Chemical, Technical Bulletin, NEODOL: Suggested Formulation, Nos. 4 and 5.

General-Purpose Cleaner, *Description: Industrial high alkaline powders. Use level: 1–4 oz./gal.*

	A	B
Nonylphenol, 9 moles ethylene oxide	5.0	10.0
Sodium carbonate (light density)	40.0	45.0
Sodium tripolyphosphate	20.0	
Sodium metasilicate, pentahydrate	35.0	
Sodium metasilicate, anhydrous		30.0
Sodium hydroxide (beads)		15.0

Source: BASF, Technical Bulletin: Suggested Formulations.

General-Purpose Cleaner, *Description: Consumer and institutional alkaline liquid. Use level: 2–8 oz./gal.*

	A	B
Water	78.1	78.7
Sodium metasilicate, anhydrous	2.4	1.0
Tetrapotassium pyrophosphate (60%)	3.0	3.0
Phosphate ester	3.0	
Sodium xylene sulfonate (40%)		3.5
Dipropylene glycol monomethyl ether	6.0	
Ethylene glycol *n*-butyl ether		6.0
Octylphenol, 9–10 moles ethylene oxide	5.0	
Octylphenol, 5 moles ethylene oxide	2.5	
C_{12}–C_{16} linear alcohol, 6.5 moles ethylene oxide		5.0
C_{12}–C_{16} linear alcohol, 3 moles ethylene oxide		2.5
Pine oil		0.3

Source: PQ Formulary Bulletin: Suggested Formulations.

General-Purpose Cleaner, *Description: Consumer and institutional nonalkaline liquid. Use level: 1–4 oz./gal.*

Water	44.5
C_9–C_{11} linear alcohol, 6 moles ethylene oxide	25.0
Sodium alkylarylsulfonate (100%)	12.0
Nonionic amide[1]	15.0
Propylene glycol	3.5

[1] 2:1 modified coconut diethanolamide (100%)
Source: Shell Chemical, Technical Bulletin, NEODOL: Suggested Formulation, Nos. 4 and 5.

General-Purpose Cleaner, *Description: Industrial high alkaline liquid. Use level: 1–4 oz./gal. Application: Can be used as a low-foaming spray cleaner.*

Water	32.2
Amphoteric surfactant[1]	5.3
Caustic potash (45%)	44.5
Tetrapotassium pyrophosphate	18.0

1 Dicarboxylic caprylic derivative, sodium salt.

Source: Rhône-Poulenc (formerly Miranol), Technical and Product Development Data: Suggested Formulation.

General-Purpose Cleaner, *Description: Industrial high alkaline powders with pine oil. Use level: 1–4 oz./gal.)*

	A	B
Sodium metasilicate, pentahydrate		55.0
Sodium orthosilicate	40.0	
Sodium carbonate (light density)	29.0	26.0
Trisodium phosphate	20.0	
Sodium tripolyphosphate		10.0
Nonylphenol, 9–10 moles ethylene oxide		6.0
Sodium alkylarylsulfonate (powder)	9.0	
Pine oil	2.0	3.0

Source: PQ Formulary Bulletin: Suggested Formulations.

General-Purpose Cleaner, *Description: Industrial liquid with pine oil and kerosene. Use level: 1–4 oz./gal.*

Water	46.65
Sodium hydroxide (100%)	0.65
Sodium tripolyphosphate	2.00
Tall oil fatty acid	5.00
Nonylphenol, 9–10 moles ethylene oxide	3.70
Pine oil	15.00
Deodorized kerosene	27.00

Source: Hercules, Pine Oil Formulary: Suggested Formulation A-3.

General-Purpose Cleaner, *Description: Consumer and institutional pine oil based liquid. Use level: 2–8 oz./gal.*

Water	64.0
Pine oil	10.0
Modified coconut polydiethanolamide	7.5
Isopropanol	17.5
Caustic potash (45%)	1.0

Source: Cyclo Chemicals, Technical Information: Suggested Formula 6-0505.

7.3   Dishwash Detergents

Dishwash applications deal with food soil removal from the various dishware hard surfaces and can be separated into two basic types, hand and machine. Both categories are designed to remove high levels of soils, which include fatty acids and esters, proteins, and carbohydrates, from dishes and glassware, eating utensils, and pots and pans. However, hand and machine dishwash applications and detergent compositions do differ significantly. Sections

7.3.1 and 7.3.2 discuss these differences as well as variations in formulations for consumer and institutional applications.

7.3.1   Hand Dishwash Detergents

Hand dishwash detergents are made for both consumer and institutional markets. Products come primarily in liquid form, although powdered hand dishwash detergents are available for household and institutional use. Consumer and institutional formulas are similar, but the consumer brands tend to use surfactants and other additives that are gentle to skin.

Liquid hand dishwash formulas are usually neutral in pH (6–9) to avoid excessive skin irritation, which can occur when sebum, the skin's fatty lubricant, is neutralized and made soluble by an alkaline solution with a high pH. This reaction defats the skin, leaving it very dry.

Liquid hand dishwash compositions usually contain high concentrations of surfactants, which are primarily anionic. Very few, if any, alkaline ingredients are incorporated. Because alkalinity is minimized, liquid hand dishwash cleaning involves very little soil modification (i.e., neutralization) to aid in the removal of fats and proteins. Consequently, soils are removed by means of the grease dissolving and emulsifying ability of the surfactant system, in addition to the mechanical action of scrubbing.

Common anionic surfactants used in liquid hand dishwash applications include alkylarylsulfonates, alkyl sulfates (AS), and alkyl ether sulfates. Alkylarylsulfonates are very cost-effective at removing oily food soils. AS and alkyl ether sulfate surfactants are milder to skin and less sensitive to water hardness than alkylarylsulfonates, although they are generally more costly. Proper combinations of these surfactants can produce a product that offers good soil removal yet is mild to skin. Among the hand dishwash formulations in Sections 7.3.1–7.3.1.5 are several examples of these surfactant mixtures.

Nonionic surface active agents, such as the ethoxylated linear alcohols and the ethoxylated alkylphenols, are also used in liquid dishwash appications but to a lesser degree than anionic surfactants. This is due to higher cost and product stability limitations associated with these surfactants. Other nonionic surfactants, such as the alkylolamides [8] and alkylamine oxides, are also incorporated to help build and stabilize suds during washing as well as to impart some mildness to skin.

Powdered hand dishwash detergents are usually alkaline products containing polyphosphates, carbonates, and anionic surfactants. They tend to be used more widely in institutional applications, such as restaurant kitchens, than by consumers. Two reasons for this trend are:

- Alkalinity is required to neutralize the heavier acidic soil loads encountered in institutional applications. This is important for maintaining the use life of the dishwash solution.
- In institutional applications, mildness to skin is not as critical as with consumer usage. Generally, employees use gloves and proper clothing to protect themselves from the corrosiveness of an alkaline dishwash solution.

Because of the alkalinity of hand dishwash products, corrosion inhibitors (e.g., aluminum compounds and silicates) to protect glass and aluminum surfaces are incorporated. The use of corrosion inhibitors is especially necessary when aluminum pots and pans are soaked for easy cleaning. Aluminum is very reactive to hydroxide ions, which are present in alkaline solutions. This reaction generates explosive hydrogen gas as a by-product. Below are some examples of typical consumer and institutional dishwash compositions (compositions indicated in weight %):

Hand Dishwash Detergent, *Description: Consumer liquids. Mixing instructions: Adjust pH to 6.5–67.0 with NaOH or mono- or triethanolamine.*

	Quality	
	Good	Premium
Sodium C$_{12}$–C$_{15}$ alcohol sulfate, 3 moles ethylene oxide (60%)	12.5	22.5
Sodium alkylarylsulfonate (60%)	25.0	27.8
Fatty acid diethanolamide	2.5	4.5
Ethanol	3.0	
Water, dye, perfume, preservatives	57.0	45.2

Source: Shell Chemical Company Detergent Formulary.

Hand Dishwash Detergent, *Description: Institutional liquid. Mixing instructions: Adjust pH to 8.5–9.0 with mono- or triethanolamine.*

Sodium alkylarylsulfonate	20.0
Octylphenol, 12–13 moles ethylene oxide	10.0
Diethanolamide of 70% lauric acid/30% myristic acid	5.0
Water	65.0

Source: Michael and Irene Ash, *A Formulary of Detergents and Other Cleaning Agents,* Chemical Publishing Company, New York, 1980.

Hand Dishwash Detergent, *Description: Institutional liquid. Mixing instructions: Adjust pH to 8.5–9.0 with mono- or triethanolamine.*

Sodium C$_{12}$–C$_{15}$ alcohol sulfate, 3 moles ethylene oxide (60%)	5.0
Sodium alkylarylsulfonate (60%)	16.7
C$_9$–C$_{11}$ linear alchohol, 8 moles ethylene oxide	3.0
Fatty acid diethanolamide	2.0
Ammonium chloride	0.2
Water	73.1

Source: Shell Chemical Company Detergent Formulary.

Hand Dishwash Detergent, *Description: Consumer and institutional low alkaline powder. Note: Replace phosphate with bicarbonate if nonphosphate product is desired.*

Sodium alkylarylsulfonate (40% flake)	25.0
Coconut fatty acid alkanolamide	3.0
Sodium tripolyphosphate	32.0
Sodium sesquicarbonate	40.0

Source: Michael and Irene Ash, *A Formulary of Detergents and Other Cleaning Agents,* Chemical Publishing Company, New York, 1980.

Hand Dishwash Detergent, *Description: Institutional, moderately alkaline powder. Note: Replace phosphate with bicarbonate and sodium citrate if nonphosphate product is desired.*

Sodium dodecyl benzenesulfonate flake	10.8
Sodium carbonate (light, granular)	19.6
Sodium tripolyphosphate	14.0
Sodium sesquicarbonate	22.0
Sodium hexametaphosphate	8.0
Sodium silicoaluminate	2.0
Sodium sulfate	17.6
Sodium chloride	6.0

Source: Michael and Irene Ash, *A Formulary of Detergents and Other Cleaning Agents,* Chemical Publishing Company, New York, 1980.

7.3.2   Machine Dishwash Detergents

Machine dishwash detergents are moderately to highly alkaline products that essentially depend on their alkalinity and oxidizing agents for soil solubilization and removal. In contrast, hand dishwash products depend primarily on surfactants for detergency.

Generally, machine dishwash detergents contain high concentrations of alkali metal polyphosphates and pyrophosphate for water hardness sequestration and soil peptization. Soluble silicates are incorporated for corrosion protection of china, glass, and metal parts, as well as for added alkalinity and soil dispersion. A chlorine bleach is incorporated to promote free rinsing and to reduce water spots [9]. Low levels of surfactants are frequently used as wetting agents to initiate soil removal. More recently, polycarboxylates have been incorporated into machine dishwash detergents at low concentrations to enhance soil suspension and rinsing of glass surfaces (5,000–70,000 MW). In some products, sodium or potassium hydroxide is incorporated as an alkaline booster to increase acid soil neutralization, especially for institutional machine dishwash applications.

Mechanical action, which is very important, is achieved by high pressure sprayed jets of water or by the whipping action of a rapidly revolving propeller [5]. In addition to mechanical cleaning, machine dishwash products rely heavily on soil neutralization to solubilize fatty acids and proteins. Because foam is generated by these neutralized soils, defoaming surfactants are often incorporated. Too much foam will inhibit the mechanical action of the water spray in the dishwasher, resulting in poor cleaning.

Machine dishwash detergents can be divided into two basic categories:

- *Automatic dishwash detergents(ADDs):* consumer products designed for use in household automatic dishwashers.
- *Machine dishwash detergents (MDDs):* commercial or institutional products designed for use in large restaurant/cafeteria machine dishwashers.

7.3.2.1   Automatic Dishwash Detergents

Consumer automatic dishwash detergents (ADDs) are lower in alkalinity than their institutional counterparts. This is, of course, for reasons of consumer safety. ADDs contain alkaline poly- and pyrophosphates for water softening and for soil suspension and peptization. In the United States, where some states have enacted phosphate bans, ADDs are exempted from mandatory removal of these phosphates because of their significant contribution to the detergent's cleaning performance. To date, there is no proven satisfactory phosphate replacement for ADD applications in the United States. However, patent literature suggests that there is a considerable amount of research under way to achieve performance parity with various nonphosphate detergent builder systems. These tend to be based on the water-softening effectiveness of citrates, zeolite A aluminosilicates, and carbonates, in conjunction with the soil-suspending properties of polycarboxylates and polysilicates. It will not be long before nonphosphate ADDs are available, using the proper combinations of these alternative builders. For the present, however, poly- and pyrophosphates will continue to be used to maintain the high level of soil removal expected from ADD compositions.

An alkaline salt commonly used in ADDs is sodium carbonate (or potassium carbonate for liquids). Sodium carbonate functions as an alkaline buffer and water softener. Although sodium carbonate is an inexpensive detergent builder, it does form insoluble calcium and magnesium carbonate scale, which can deposit inside the dishwasher, adhering to the walls, clogging spray openings, and so on. To make a good dishwash product, care should be taken to use the proper amount of sequestering agents, in combination with carbonates, to prevent scale buildup.

Alkaline sodium and/or potassium silicates are incorporated into ADDs for corrosion protection of china overglaze and metal surfaces, as well as for soil neutralization and dispersion. Typical weight ratios of SiO_2 to Na_2O used in consumer ADD applications range from 1.6 to 2.88. In general, the higher the silica-to-alkali ratio, the better the corrosion protection [10]. For powdered ADD products, a 2.0–2.4 weight ratio appears to be optimal. Between these ratios, sodium silicate exhibits the best solubility rate for an ADD application at the highest silica-to-alkali ratio. Other corrosion inhibitors include various aluminum compounds, such as sodium aluminate and sodium silicoaluminate [11].

A chlorine bleach source, such as dichloroisocyanurate or chlorinated trisodium phosphate (for powders) and sodium hypchlorite (for liquids), is incorporated to help promote free rinsing and to reduce spotting. Chlorine is very effective at breaking down protein to soluble components, which are then removed by mechanical and chemical cleaning processes. Without chlorine, small food particles would cling to dishes and glassware, allowing drops of water to remain through the rinse cycle. When these droplets dry, they leave residual solids that appear as spots [12]. Available chlorine will also sanitize china, glass, and utensil surfaces and bleach out stains. Because of chlorine's negative environmental impact (formulation of chlorinated organic compounds), other oxidizing bleach systems, such as perborate and percarbonate, are under investigation. Although presently there are no major U.S. consumer brands using nonchlorine bleaches, the patent literature suggests that in the near future more oxygen bleaches may be seen on the market.

Low-foaming surface active agents, both anionic and nonionic, are used as wetting agents to help initiate soil removal, and as rinse aids to improve the sheeting action of water. The surfactant's role in automatic dishwash detergents is minimal in comparison to hand dishwash applications. However, the surfactant will add significantly to an ADD's performance and is therefore necessary. When choosing a surfactant for liquid ADD formulations, it is imporant to keep in mind the need for stability in the presence of chlorine. Otherwise the surfactant molecule will degrade and the available chlorine will diminish. Also, as mentioned earlier, to minimize foam generated during the cleaning cycle, a defoaming surfactant (e.g., phosphate ester) may be incorporated [13].

Also used in ADD formulations are polycarboxylates, such as polyacrylic and acrylic/maleic acid copolymers with molecular weights between 5,000 and 70,000. These synthetic polymers are effective at sequestering water hardness and preventing scale formation. Higher molecular weight polymers will help to disperse and suspend soils, as well as to improve the sheeting action of water when rinsing [14].

The main objectives when developing a powdered or liquid ADD, apart from good detergency, is the final product's density and ease of dispensing. Consumer dishwashers have small dispensing cups that sit vertically when the door is shut. To be sure that enough detergent can fit into a small volume, the ADD's density must be relatively high. When developing a liquid ADD, the need to avoid leakage from the cup before the wash cycle begins dictates modification of the product's rheology by increasing its viscosity. To meet these product-dispensing criteria, typical automatic dishwash detergents are manufactured as dense, agglomerated, granular products or as concentrated liquid slurries and gels.

Dense dry-blended ADD powders are also manufactured and are very similar to the agglomerated products in composition. They differ only in method of manufacture, level of hydration (agglomerated products tend to be more hydrated for quicker solubility), and source of silicate. When making agglomerated ADDs, liquid silicate is typically used, not only for corrosion protection and detergency, but as a processing aid to bind or glue ingredients into granules. However, dry-blended ADDs tend to incorporate powdered forms of sodium silicate for ease of handling and blending. Below are examples of typical ADD compositions, both powders and liquids (compositions indicated in weight %):

Automatic Dishwash Detergent, *Description: Consumer agglomerated powder.*

Sodium tripolyphosphate	34.8
Sodium carbonate	19.0
Low-foaming nonionic surfactant	3.0
Sodium dichloroisocyanurate, $2H_2O$	1.8
Sodium sulfate	21.5
Moisture	7.9
Liquid sodium silicate[1]	12.0

[1] 2.4 weight ratio, 42–47% solids; used as a binder for agglomeration as well as for corrosion protection of china and glass.
Source: Colgate-Palmolive, U.S. Patent application 444 250, Jan. 12, 1989.

Automatic Dishwash Detergent, *Description: Consumer dry blended powder.*

	Water hardness	
	Soft-medium	Hard
Sodium tripolyphosphate	20.0	30.0
Sodium tripolyphosphate, $6H_2O$	8.0	15.0
Sodium hydrous polysilicate[1]	25.0	25.0
Sodium carbonate	15.0	10.0
Low-foaming nonionic surfactant	2.0	3.0
Sodium polyacrylate, MW 4,500)	1.0	2.0
Sodium dichloroisocyanurate, $2H_2O$	2.0	2.0
Sodium sulfate	27.0	13.0

[1] 2.0 weight ratio powder with 17.5% water.
Source: PQ Detergent Formulary.

Automatic Dishwash Detergent, *Description: Consumer dry blended nonphosphated powder.*

	A	B	C
Sodium citrate	15.0	24.0	
Sodium hydrous polysilicate[1]	20.0	20.0	20.0
Sodium carbonate	25.0	25.0	30.0
Low-foaming surfactant	3.0	3.0	3.0
Acrylic/maleic acid copolymer[2]	2.0	2.0	2.0
Sodium dichloroisocyanurate	2.0	2.0	2.0
Ethylene diaminetetraacetic acid			18.0
Zeolite A aluminosilicate	15.0		
Sodium sulfate	18.0	24.0	25.0

[1] 2.0 weight ratio powder with 17.5% water.
[2] Sodium salt; MW 70,000

Source: PQ Corporation Detergent Formulary.

Automatic Dishwash Detergent, *Description: Consumer liquid slurry. Note: If product instability occurs in the form of crystallization, use potassium salts in place of sodium (e.g., potassium silicate for sodium silicate).*

Sodium tripolyphosphate	30.0
Sodium hydroxide (50%)	20.0
Liquid sodium silicate[1]	20.0
Sodium n-decyl diphenyloxide disulfonate (45%)	2.0
Sodium hypochlorite (13% available chlorine)	15.0
Sodium bentonite	1–3
Water	Balance

[1] 2.4 weight ratio, 47.5% solids.
Source: BASF Conferences on Autodish Technology, Sept. 22, 1987.

Automatic Dishwash Detergent, *Description: Consumer liquid gel.*

Polyacrylic acid polymer[1]	0.9
Tetrapotassium pyrophosphate	15.0
Sodium tripolyphosphate $6H_2O$	13.0
Sodium hydroxide (50%)	2.4
Liquid sodium silicate[2]	21.0
Phosphate ester (defoamer)	3.2
Sodium n-decyl diphenyloxide disulfonate (45%)	1.0
Sodium hypochlorite (13% avaoilable chlorine)	7.5
Water	36.0

[1] High molecular weight thickener.
[2] 2.4 weight ratio, 47.5% solids.
Source: Colgate-Palmolive, U.S. Patent application 353 712, May 18, 1989.

7.3.2.2   Machine Dishwash Detergents

Institutional machine dishwash detergents are similar to the consumer ADD formulations but higher in alkalinity. High alkalinity is necessary in institutional dishwash applications because of the increased acidic soil loads. To achieve high active alkalinity, powdered formulations usually include high concentrations of sodium metasilicate as well as trisodium orthophosphate. Often, sodium hydroxide beads are included as an alkaline booster. The high alkalinity of these detergents mandates the presence of a corrosion inhibitor to prevent etching of glass, china, and metal surfaces. Sodium metasilicate is effective here.

Liquid machine dishwash detergents are typically formulated as highly concentrated alkaline solutions containing mixed sodium and potassium salts. Potassium salts increase the final product's solubility. This is important for good shelf life, especially when the finished liquid product is stored under cold weather conditions [15]. Liquid machine dishwash products often contain sodium hypochlorite bleach as well as low levels of wetting agents. When choosing a wetting agent for a machine dishwash liquid, the following surfactant properties are important:

- Good solubility in highly concentrated alkaline solutions.
- Good stability in the presence of chlorine bleach if hypochlorite is used.

Typically, institutional machine dishwashers run through their wash cycles faster and at higher temperatures than home dishwash units. Machine dishwash systems can range from batch units to large conveyor systems that run hundreds of dishes per hour. Also, many institutional machine dishwash systems use computer-driven peristatic pumps to feed liquid detergent products through plastic tubing into the dishwashing process. This automated approach takes these high alkaline products out of the hands of operators and allows them to be dispensed in an accurate, precise, and timely fashion. This technology has been primarily responsible for the growth of liquid machine dishwash products in the last decade. Below are several examples of powdered and liquid formulations for institutional machine dishwash applications (compositions indicated in weight %):

Machine Dishwash Detergent, *Description: Institutional moderately alkaline powder.*

Sodium tripolyphosphate	40.0
Trisodium phosphate, chlorinated	30.0
Sodium metasilicate, anhydrous	28.0
Nonionic surfactant[1]	2.0

[1] Alkoxylated biodegradable hydrophobe, low foaming.
Source: Olin Chemicals, Technical Bulletin, Chemicals for Cleaning.

Machine Dishwash Detergent, *Description: Institutional high alkaline powder.*

Sodium tripolyphosphate	40.0
Sodium hydroxide	20.0
Sodium metasilicate, anhydrous	20.0
Dichloroisocyanurate, $2H_2O$	4.0
Sodium carbonate	14–16
Defoaming surfactant	0–2

Source: R. Maldonado, *Phosphates for Innovative Product Forms,* FMC Corporation, March 26, 1991.

Machine Dishwash Detergent, *Description: Institutional nonphosphated powder.*

Sodium metasilicate, pentahydrate	55.3
Nitrilotriacetic acid, sodium salt	41.0
Sodium chloroisocyanurate	2.0
Low-foaming surfactant	0.7
Poly(ethylene glycol)	1.0

Source: PQ Corporation Detergent Formulary.

Machine Dishwash Detergent, *Description: Institutional alkaline liquid.*

Tetrapotassium pyrophosphate (60%)	23.3
Potassium hydroxide (45%)	9.0
Liquid potassium silicate[1]	27.0
Potassium carbonate	12.0
Amphoteric surfactant[2]	2.0
Water	26.7

[1] 2.5 weight ratio, 29% solids.
[2] Low-foaming sodium salt.
Source: Rhône-Poulenc (formerly Miranol Chemical), Technical Data, Amphoteric Imidazoline: Suggested Formulation.

Machine Dishwash Detergent, *Description: Institutional high alkaline liquid (no wetting agent present).*

Tetrapotassium pyrophosphate (60%)	20.0
Potassium hydroxide (45%)	35.0
Liquid potassium silicate[1]	30.0
Sodium hypochlorite (15% available chlorine)	5–10
Water	Balance

[1] 2.5 weight ratio, 29% solids.
Source: PQ Corporation Detergent Formulary.

Machine Dishwash Detergent, *Description: Institutional nonphosphated alkaline liquid.*

Potassium hydroxide (45%)	20.0
Liquid potassium silicate[1]	22.0
Amphoteric surfactant[2]	1.0
Gluconic acid	8.0
Water	49.0

[1] 2.5 ratio, 29% solids.
[2] Low-foaming sodium salt.
Source: Rhône-Poulenc (formerly Miranol Chemical), Technical Data, Amphoteric Imidazoline: Suggested Formulation.

Machine Dishwash Detergent, *Description: Institutional high viscosity (ca. 10^3 cP) alkaline liquid. Mixing instructions: Dissolve vinyl ethyl/maleic anhydride copolymer in nonylphenol/water mixture by stirring at room temperature for 8 hours. Heat, if necessary, to complete dissolution of copolymer. Follow by addition of remaining ingredients with mixing.*

Nonylphenol, 15 moles ethylene oxide	0.5
Vinyl ethyl/maleic anhydride copolymer	1.0
Potassium hydroxide (50%)	3.0
Liquid sodium silicate[1]	20.6
Low-foaming nonionic surfactant[2]	3.0
Tetrapotassium pyrophosphate (60%)	63.6
Water	8.6

[1] 1.8 weight ratio, 38% solids, clear grade.
[2] Modified linear aliphatic polyether, 95% concentrate.
Source: Rhône-Poulenc (formerly GAF Surfactants) Formulary: Prototype Formulation GAF 5233.

7.4   Vehicle Wash Cleaners

This section deals with the industrial cleaning of car and truck hard surfaces. Many owners of cars and small trucks wash their vehicles at home, using heavy-duty, general-purpose cleaners. These cleaners, examples of which can be found in Section 7.2.1, are very effective at removing the grease and particulate soils typically present. However, there is a large vehicle wash industry which uses a variety of cleaning products in applications ranging from local car washes to commercial transportation wash operations specializing in cleaning buses and large trucks.

Industrial vehicle wash processes are usually designed as a high pressure spray (HSP) system or a combination of spray and mechanical brush. High pressure spray cleaning, as in metal and machine dishwash applications, provides an intense, high energy method to physically remove soils from vehicles with the aid of a cleaning product. The wash solution is usually hot and contains low concentrations of a moderately to highly alkaline cleaner with minimal foaming characteristics.

A vehicle wash system consisting of spray and mechanical brush, common at local car washes, is also effective at soil removal. However, there is less mechanical energy used and more reliance on the detersive properties of the cleaner. Generally, the vehicle wash cleaner is formulated with increased levels of surfactants that foam and is used at higher concentrations than in HSP systems.

The soils commonly found on vehicles range from petroleum-based oils and greases, to clay and carbon particulates, to slightly acidic soils such as bird droppings and tree sap. The soil load for this application can be high, especially with commercial vehicles (e.g., tractor trailer trucks and tankers). Therefore, to handle the heavy soils of these applications, vehicle wash cleaners are characteristically formulated at high concentrations of active ingredients. These ingredients include surfactants, alkaline builders, organic sequestering agents, dispersing agents and, in some cases, solvents.

Surfactants, both anionic and nonionic, are used to wet the vehicle surface, initiate removal of particulates, and emulsify oily soils. Examples are the alkylarylsulfonates and alkylphenol ethoxylates, respectively. Water miscible solvents, such as the glycol ethers, aid in grease removal. However, these solvents may harm paint, plastic, and rubber parts.

Very important to a vehicle wash cleaner's performance is the alkaline system, which is usually comprised of STPP, tetrasodium pyrophosphate (TSPP), and/or trisodium phosphate (TSP), or their potassium analogs, at high concentrations. STPP and TSPP will sequester calcium and magnesium ions, often present at higher than normal levels due to road grime. These water hardness ions, if not removed, will diminish the cleaning effectiveness of the surfactant system. TSP and sodium carbonate, both good sources of alkalinity, are also used to remove calcium and magnesium via precipitation. Other builders commonly used include organic chelating agents, such as sodium gluconate, sodium citrate, EDTA, and phosphonates. These organic builders are frequently used in liquid car and truck wash cleaners because they have excellent stability with surfactants.

The vehicle wash industry deals with many hard surfaces of different types, ranging from paint and lacquer overcoats to aluminum alloys and glass. In many cases, these surfaces are easily corroded or etched when subjected to highly alkaline solutions. Aluminum is particularly sensitive to chemical corrosion, which is exhibited by a bluish cast referred to as "aluminum blush." Therefore, it is important to design a cleaner that not only removes soil effectively but also minimizes interaction with these substrates. When cleaning under moderately to highly alkaline conditions, the use of sodium silicate is an excellent way to provide this corrosion protection to aluminum, paint, and glass surfaces, as well as to add alkalinity and increased soil dispersion activity [16].

Over time, surface oxidation will cause the aluminum parts of vehicles to dull. To brighten these parts, hydrofluoric acid and/or bifluoride salts as well as phosphoric acid are commonly used. Such treatment removes both the oxide and road film. However, fluoride compounds are hazardous and should be handled very carefully. If an alternative to hydrofluoric acid or fluoride salt is desired, sodium nitrilotriacetate monohydrate may be used to produce a similar effect [17].

Anodized aluminum parts, used as trim, bumpers, and wheel covers, are commonly found on modern cars and commercial vehicles. Anodized aluminum has a surface film formed by immersion in a sulfuric acid solution, after which an electric current is applied, using the aluminmum as the anode. A protective oxide film is formed and reacted with nickel acetate. This amphoteric anodic coating is readily dissolved by acidic solutions below a pH of 4 and by most alkaline solutions above pH of 10. Corrosion mechanisms commonly involve film dissolution by hydroxyl ions (alkali attack) and removal of nickel from the coating by sequestering agents (chelate blush). To prevent alkali attack, the minimum weight ratio of SiO_2 to active Na_2O in the final formulation should be 0.80. Inhibition of "chelate blush" is more complicated. The minimum weight ratio of SiO_2 to chelator should be 4.0, 4.2, 8.2, and 16.0 parts of SiO_2 per 100.0 parts of chelator when phosphate glass, sodium tripolyphosphate, tetrasodium pyrophosphate, and organic chelators (e.g., citrate, NTA, and EDTA) are used, respectively [18].

Below are examples of vehicle wash formulations (compositions indicated in weight %):

Car and Truck Wash Detergent, *Description: Commercial alkaline liquid. Use level: 3–9 oz./gal.*

Tetrapotassium pyrophosphate (60%)	15–20
Liquid sodium silicate[1]	12–14
Sodium xylene sulfonate	10–15
C_{12}–C_{15} linear alcohol, 9 moles ethylene oxide	5.0
Water	Balance

[1] 1.8 weight ratio liquid, 38% solids, clear grade.
Source: FMC Corporation, Technical Data: Detergent Applications Bulletin No. 4.

Car and Truck Wash Detergents, *Description: Commercial nonphosphated alkaline liquid. Use level: 3–9 oz./gal.*

Sodium metasilicate, anhydrous	1.0
Ethylenediaminetetraacetic acid, sodium salt (40%)	19.3
Phosphate Ester[1]	2.0
Sodium hydroxide (50%)	0.7
Nonionic/amphoteric surfactant[2]	3.0
Water	74.0

[1] Low-foaming surfactant hydrotrope.
[2] Imidazoline amphoteric.
Source: Mona Industries, Technical Bulletin: Subject—Monafax 1293, April 1988.

Car and Truck Wash Detergents, *Descriptions: Commercial nonphosphated alkaline liquid; aluminum brightener. Use level: 13.5 oz./gal.*

Ethylenediaminetetraacetic acid	10.0
Liquid sodium silicate[1]	14.8
Sodium xylene sulfonate (40%)	7.5
C_9–C_{11} linear alcohol, 6 moles ethylene oxide	3.0
Methylene phosphate, Na_5 amino Tris	1.0
Water	63.7

[1] 2.5 weight ratio liquid, 38% solids, clear grade.
Source: Monsanto, Technical Data: Suggested Formulations.

Car and Truck Wash Detergents, *Description: Commercial alkaline powder. Application: Dissolve in stock tank and meter into wash system. Composition B has lower foaming surfactant system and can be used in high pressure spray applications. Use level: 0.5–4 oz./gal. Note: To convert both formulations to a nonphosphate status, use 25–30% EDTA in place of the phosphate compounds. To increase soil dispersion and prevent precipitation of calcium and magnesium salts, polycarboxylates, such as acrylic/maleic acid copolymers with molecular weights between 4,500 and 20,000 can be used at 1–3% solids. The balance of the replacement can be sodium carbonate and/or sodium sulfate.*

	A	B
Sodium metasilicate, anhydrous	20.0	
Sodium metasilicate, pentahydrate		30.0
Soda ash (light density)	20.0	10.0
Sodium tripolyphosphate	40.0	44.0
Trisodium orthophosphate	5.0	
Sodium alkylarylsulfonate (90%)	5.0	5.0
C_9–C_{11} linear alcohol, 6 moles ethylene oxide	10.0	
Octylphenol, 7–8 moles ethylene oxide		5.0
Octylphenol, 5 moles ethylene oxide		5.0
Carboxymethylcellulose		1.0

Source: PQ Corporation Detergent Formulary.

Car and Truck Wash Detergents, *Description: Commercial alkaline powder; aluminum brightener. Use level: 2.0 oz./gal.*

Ethylenediaminetetraacetic acid, Na_4	10.0
Sodium tripolyphosphate	68.0
Sodium silicate[1]	12.0
C_9–C_{11} linear alcohol, 6 moles ethylene oxide	10.0

[1] 2.0 weight ratio powder, 82.5% solids.
Source: Monsanto, Technical Data: Suggested Formulations.

Truck Wash Detergents, *Description: Commercial alkaline powder. Application: Dissolve in stock tank and meter into wash system. Use level: 1–3 oz./gal. (light to moderate soil); 4 oz./gal. (heavy soil).*

	A	B
Sodium metasilicate, pentahydrate	20.0	20.0
Soda ash (light density)	40.0	30.0
Sodium tripolyphosphate		35.0
Tetrasodium pyrophosphate	25.0	
Sodium alkylarylsulfonate flake (90%)	10.0	15.0
Nonylphenol, 9–10 moles ethylene oxide	5.0	

Source: PQ Corporation Detergent Formulary.

Vehicle Wash Detergents, *Description: Commercial alkaline powder; safe for anodized aluminum.*

	A	B
Sodium metasilicate, anhydrous	9.0	
Sodium silicate[1]		7.0
Soda ash (light density)		5.0
Sodium tripolyphosphate	40.0	40.0
Tetrasodium pyrophosphate	10.0	10.0
Sodium alkylarylsulfonate flake (90%)	10.0	10.0
Nonylphenol, 9–10 moles ethylene oxide	3.0	3.0
Sodium sulfate	28.0	25.0
Weight % SiO_2/active Na_2O	1.12	0.81
Parts SiO_2/parts chelator	8/100	8/100

[1] 2.4 weight ratio powder, 82.5% solids.
Source: FMC Corporation, Detergent Application Bulletin No. 8: Guideline Calculations for Alkaline Vehicle Detergents.

Aluminum Truck Body Cleaner–Brightener, *Description: Commercial acidic liquid.*

	A	B	C
Phophoric acid (85%)	35–40	10	47.2
Nonionic surfactant[1]	5–10	4	2.0
Glycol ether[2]			16.0
Sodium bifluoride	1–2		
Hydrofluoric acid (48%)		15	
Water	Balance	Balance	34.8

[1] Octylphenol or linear alcohol, 7–9 moles ethylene oxide.
[2] Ethylene glycol monobutyl ether.
Source: FMC Corporation, Technical Data: Detergent Applications Bulletin No. 4.

7.5   Metal Cleaners

This section addresses hard-surface cleaning systems designed to remove soil from metal substrates. When preparing metals for finishing, the most important consideration by far is the cleaning process. The appearance and acceptance of the finished product depends primarily on a good foundation for the finish, which is achieved with a clean and active substrate [19].

Some of the major metal industries processes that require cleaning include the following [5]:

- Cleaning before electroplating
- Preparation for application of paint-bonding processes
- Preparation for porcelainizing or ceramic coatings
- Removal of residues that could cause corrosion or interfere with processing as in drawn tube, wire, or rod; fabricated assemblies; and rolled strip
- Preparation of chemical surface coatings (anodizing, black oxide, and electropolished metals such as aluminum, copper, and stainless steel)
- Preparation for hot dip coatings (galvanized, etc.)
- Cleaning scrap metal before melting.

The variety of soils encountered in metal cleaning is extensive. Generally, they are comprised of fatty greases or mineral oils used as lubricants; soot, pigments, and other particulate soils from upstream processes; and metal oxides, corrosion inhibitors, and drawing compounds. To remove these tenacious soils thoroughly, metal surface preparation requires highly active cleaning systems. These cleaning systems involve formulated products containing solvents, surfactants, alkalies, or acids alone or in various combinations, as well as high temperatures (60–90 °C), mechanical energy (high pressure spray and/or brushing, and electrophysical energy (electrolytic cleaning).

In metal cleaning, as in other cleaning applications, the cleaner composition used for soil removal depends largely on the type of soil, especially its physical and chemical properties. For example, when cold rolled steel has become coated with tallow, immersion in a hot, highly alkaline cleaner solution followed by brushing is very effective: it removes this fatty lubricant by converting it to a soluble soap. Also important in the selection of a cleaning system is the soil's condition after it has undergone chemical changes through aging, drying out, or heating during metal working. These chemical changes, such as oil polymerization or metal soap formation, may have produced soils which are very difficult to remove. A series of cleaning steps designed to soften and break up the soil before removal is required.

7.5.1   Cleaning Systems

Most metal cleaning involves one or a combination of the following processes: soak or immersion, spray, and electrolytic.

Immersion and spray cleaning processes are commonly used with the following cleaning systems [20]:

1. *Solvent cleaning,* which involves the use of petroleum or chlorinated solvents that function by dissolving all or part of the soil. These solvents are used with surfactants that render them emulsifiable in a water rinse.
2. *Emulsion cleaning,* which involves the use of oil-in-water emulsions, which are usually alkaline (pH 7.8–10.0) and relatively hot (60–80 °C). Cleaning is accomplished by dissolving and/or emulsifying the soil.
3. *Acid cleaning,* which involves the use of inorganic and organic acids (e.g., phosphoric, sulfuric, gluconic, acetic) along with water-miscible solvents and surfactants. Acid cleaning removes soils by dissolving metal oxides and by emulsifying oils.
4. *Detergent cleaning,* which involves the use of formulated detergents that incorporate alkaline buffers, sequestering agents, dispersant, corrosion inhibitors, and surfactants. These detergents function by wetting, emulsifying, dispersing, and solubilizing the soil, usually at high temperatures (50–94 °C). Use concentrations range from 6 to 12 oz./gal.
5. *High alkalinity cleaning,* which involves the use of formulated alkaline cleaners composed of sodium hydroxide, sodium metasilicate, carbonates, sequestering agents, dispersant, and various surfactants. Cleaning is usually performed at high temperatures (50–94 °C) and at concentrations ranging from 8 to 32 oz./gal. These cleaners effectively remove oils and greases, especially animal fats, as well as smut and light scale.
6. *Electrolytic cleaning,* which involves the use of heavy-duty alkaline cleaner systems with an applied electric current. In electrolytic cleaning, water is electrolyzed to oxygen at the anode and hydrogen at the cathode. The metal surface can be set up as either the anode or cathode, or one can alternate between the two as a periodic reverse current. The gas generated at the metal surface helps remove embedded soils, while reversing the current prevents deposition of any metallic film or charged contaminates. Electrocleaners, as

they are called, are designed for soil removal and surface activation to accept plating. The following formulations and applications are typical of some metal cleaning systems.

Note: A common method, called the water break test, is used to determine metal surface cleanliness as part of an evaluation to determine a cleaner's performance. This evaluation involves visual examination of the metal surface after a final rinse in cool, clean water. The criterion for passing is a continuous sheet of water on the surface, indicating complete soil removal. If the water breaks up into patches and does not wet the surface completely, the specimen fails and there is an indication that residual soils are present.

Metal Cleaner: Steel, *Description: Detergent liquid. Cleaning process: Immersion and/or spray. Use level: 3–12 oz./gal.*

Octylphenol, 9–10 moles ethylene oxide	2.0
Phosphate ester (50%)	6.0
Potassium hydroxide	12.0
Sodium metasilicate, anhydrous	12.0
Tetrapotassium pyrophosphate	12.0
Water	56.0

Source: Rohm & Haas Formulary, 1986.

Metal Cleaner: Aluminum, *Description: Detergent powder. Cleaning process: Immersion. Use dilution: 2–6 oz./gal.*

Tetrasodium pyrophosphate	30.0
Sodium metasilicate, pentahydrate	30.0
Sodium carbonate	20.0
Trisodium phosphate	17.0
Nonylphenol, 9 moles ethylene oxide	3.0

Source: Monsanto Bulletin, Metal Treatment, 1991.

Metal Cleaner: Iron and Steel, Description: Alkaline powder. Cleaning process: Immersion. Use dilution: 6–8 oz./gal.

Tetrasodium pyrophosphate	20.0
Sodium metasilicate, pentahydrate	20.0
Soda ash	53.0
Sodium gluconate	5.0
Nonylphenol, 9 moles ethylene oxide	2.0

Source: Monsanto Bulletin, Metal Treatment, 1991.

Metal Cleaner: Steel, *Description: Heavy alkaline powder. Cleaning process: Immersion, spray, or electrolytic. Use dilution: 2–4 oz./gal.*

	Phosphated	Nonphosphated
Octylphenol, 9–10 moles ethylene oxide	2.0	2.0
Sodium metasilicate, anhydrous	35.0	35.0
Tetrasodium pyrophosphate	20.0	
Sodium hydroxide, beads	18.0	32.0
Sodium carbonate	25.0	31.0

Source: PQ Corporation Detergent Formulary.

Metal Cleaner: Aluminum, *Description: Acid liquid. Cleaning process: Spray. Use dilution: 1–2 oz./ gal.*

Sulfuric acid (50%)	40.0
Amphoteric surfactant[1]	5.0
Water	55.0

[1] 2-alkylimidazoline (100% solids).
Source: Michael and Irene Ash, *Formulary of Detergents and Other Cleaning Agents,* Chemical Publishing Com-pany, New York, 1980.

Metal Cleaner: Aluminum, *Description: Acid liquid. Cleaning process: Immersion or spray. Use dilution: 2–6 oz./gal.*

Anionic surfactant[1]	5.0
Ethylene glycol *n*-butyl ether	6.0
Phosphoric acid (85%)	38.0
Hydrofluoric acid (70%)	8.0
Ethylenediaminetetraacetic acid	1.0
Water	42.0

[1] Dicarboxylic coconut derivative, disodium salt (39% solids).
Source: Michael and Irene Ash, *Formulary of Detergents and Other Cleaning Agents,* Chemical Publishing Com-pany, New York, 1980.

Metal Cleaner-Brightener: Aluminum, *Description: Alkaline–emulsifier liquid. Cleaning process: Immersion or spray. Use dilution: 1–4 oz./gal.*

Nonionic surfactant[1]	5.0
Sodium alkylnaphthalenesulfonate	3.0
Sodium metasilicate, anhydrous	3.0
Tetrapotassium pyrophosphate	3.0
Dipropylene glycol methyl ether	5.0
Water	81.0

[1] Polyglucoside (70% solids), alkali soluble.
Source: Rohm & Haas Formulary, 1986.

Aluminum Etching Bath, *Description: Alkaline liquid. Cleaning process: Immersion. Use dilution: None.*

Trisodium phosphate, crystalline	0.5
Sodium gluconate	0.2
Sodium carbonate	1.0
Hydroxyethylidene diphosphonic acid (60%)	0.2
Sodium hydroxide (50%)	7.0
Water	91.1

Source: Monsanto Bulletin, Metal Treatment, 1991.

Metal Cleaner-Brightener: Aluminum, *Description: Alkaline nonfluoride liquid. Cleaning process: Immersion or spray. Use dilution: 1–4 oz./gal.*

Trisodium phosphate	2.0
Sodium hydroxide (50%)	4.0
Gluconic acid (50%)	2.0
Chelating agent[1]	2.0
Nonylphenol, 9 moles ethylene oxide	0.5
Water	89.5

[1] Amino Tris methane phosphonic acid (50% solids).
Source: Monsanto Bulletin, Metal Treatment, 1991.

Metal Cleaner-Brightener: Aluminum, *Description: Alkaline nonfluoride liquid. Cleaning process: Immersion or spray. Use dilution: 1–4 oz./gal.*

Ethylenediaminetetraacetic acid	2.0
Sodium gluconate	2.0
Potassium hydroxide (45%)	20.0
Liquid sodium silicate, 3.22 ratio, 42 Baumé	20.0
Amphoteric alkali surfactant	3.0
Octylphenol, 9 moles ethylene oxide	2.0
Water	51.0

Source: Exxon Chemical Company, 1989 Formulary.

Metal Parts Cleaners, *Description: Solvent–emulsifier liquid. Cleaning process: Immersion. Use directions: Composition A: submerge metal parts in the solution; agitate or scrub parts. After removing them, rinse with water and dry. Composition B: paint on greasy metal parts and hose off.*

	A	B
Anionic surfactant[1]		5.0
Octylphenol, 5 moles ethylene oxide	12.0	
Kerosene	83.0	45.0
Heavy aromatic naphtha		40.0
Cresylic acid	5.0	10.0

[1] Dioctyl sodium sulfosuccinate (64%).
Source: Rohm & Haas Formulary, 1986.

References

1. Milwidsky, B.M., Household chemical specialties, HAPPI, *26* (1989).
2. Cox, M.F., Matson, T.P., J. Am. Oil Chem. Soc., *61*, 1273 (1984).
3 Falbe, J., Ed., *Surfactants in Consumer Products,* Springer-Verlag, Berlin.
4. Arcosolve Update: Glycol Ethers & Glycol Ether Acetates, Bulletin, ARCO Chemical Co.,
5. Davidsohn, A., Milwidsky, B.M., *Synthetic Detergents,* Wiley, New York, 1978.
6. Cutler, W.G., Davis, R.C., Eds., *Detergency, Theory and Test Methods,* Part II, Dekker, New York, 1975.
7. Levitt, B., *Oils, Detergents, and Maintenance Specialties,* Vol. 11, Chemical Publishing Co., New York, 1967.
8. Chalmers, L., Bathe, P., *Household and Industrial Chemical Specialties,* 2nd ed., Chemical Publishing Co., New York, 1978.
9. Fuchs, R.J., Formulation of Household Automatic Dishwash Detergents, Applications Bulletin No. 6, FMC Corp.
10. Weldes, H., Polysilicates as Detergent Builders, Bulletin, PQ Corp., Conshohocken, PA.
11. Knapp, K., Thompson, J., U.S. Patent 3,225,117; Austin, A., U.S. Patent 3,755,130.
12. Albrecht, K., Soap Chem. Spec., *31*(1), 33; *31*(2), 44 (1955).
13. Dixit, N., et al., U.S. Patent Application, 353712.
14. Polymeric Dispersing Agents, Bulletin ES-8714E, BASF.
15. Holtzer, H., Esmonde, B., Stability of high solids alkali silicates, Soap Cosmet. Chem. Spec., (Oct. 1988).
16. Falcone, J.S., Spencer, R.W., A Summary of the Role of Silicates in Detergents, Bulletin, PQ Corp., Conshohocken, PA.
17. NTA & Vehicle Wash Applications, Bulletin, Monsanto Chemicals.
18. Cohen, E.L., Hook, J.A., Corrosion of Anodized Aluminum by Alkaline Cleaners, Bulletin, FMC Corp.
19. Innes, W.P., Metal Cleaning, Metal Finishing, Bulletin, MacDermid Inc., 1991.
20. Spring, S., *Industrial Cleaning,* Prism Press, Melbourne, Australia, 1974.

CHAPTER 8

Textile Industry Applications

K. Robert Lange

Textile scours are the formulations used to free the fibers of soil to prepare them for further processing. The raw fibers must be washed before further operations (carding, spinning, weaving, knitting, etc.) can begin. These additional steps require the application of lubricants and sizes, so that further scouring is then needed to prepare the fabric for dyeing or printing.

Of the fiber sources commonly used in North America, wool requires by far the most scouring owing to its origin. Low alkalinity must be maintained for wool scouring to prevent damage to the fiber structure. On the other hand, cotton, the other major natural fiber, requires high alkalinity scouring, which swells the fibers, allowing access to the lumen and removing soil from the surface.

Synthetic fibers are generally clean when made. Scouring to remove incidental soil and lubricants can generally be accomplished using mild alkalinity and wetting agents. Rayon requires special care, avoiding strong acids and bases or solvents such as acetone and esters. Blends of synthetic and natural fibers are generally treated with respect to the natural fiber component, since this component usually has the most soil.

8.1   Introduction

Textile scouring is a term that originated with wool processing, and purists maintain that its use should be restricted to that fiber. Since the term has been used for all fibers in everyday parlance, however, it is so used here. The essentials of the process, as is the case with most industrial processes, entail both the cleaning products used and the equipment. The latter has had a tendency to be slow in developing, compared with other textile processes such as spinning. Therefore, the chemicals employed have remained relatively simple. In the future this may well change, should the trends toward natural fibers continue [1].

In the textile industry, the steps in the fabrication of finished material can be roughly classified as follows [2]:

1. Natural fibers are treated in preparation for spinning. This includes carding, a mechanical process that aligns the fibers.
2. Spinning, which forms the fibers into yarn. For synthetic fibers this is the step where the melt or solution is extruded into fibers.
3. Fabric forming, which includes knitting, weaving and felting.
4. Printing or dyeing. For synthetics, this step may precede spinning. In such cases pigment is added to the melt. Yarn bundles and fabric are generally treated with disperse dyes, in the case of polyester.

K. Robert Lange, 805 Lombard Street, Philadelphia, Pennsylvania 19147, U.S.A.

5. Finishing, where the fabric is treated with softeners, crease-resistant additives, soil-resistant chemicals, or even mechanical methods such as sanding to provide the desired appearance and feel.

Scouring is needed between the latter steps for the removal of soil or lubricants that have been applied to the goods. Very simply, scouring utilizes alkalinity with the addition of surfactants as wetting agents to clean the fibers. Some surfactants are generated in the scouring of raw fiber, as the fats present become hydrolyzed, producing soaps. Wool and silk are sensitive to alkali, tending to hydrolyze; therefore milder alkalinity is needed for them than for cotton or linen. Synthetic fibers are generally free from soil, except for applied lubricants, needing only mild washing. Blends of synthetic and natural fibers need to be treated for the most heavily soiled component, the natural fiber [3].

Since the present cleaning practices in the industry involve simple combinations of commodity chemicals, there is not much opportunity to use specialty formulations. One area of interest is the use of enzymes for the digestion of starch sizing or removal of protein residues from wool.

It is always important to understand the properties of the substrate being cleaned, if for no other reason than to prevent damage to it. This chapter outlines the important properties for the most commonly encountered fibers and representative scours in use.

8.2   Cotton

Cotton is a cellulosic product obtained from the cotton plant. As such, it consists of hollow hairs or fibers that can contain small amounts of fats in the interior lumen and lower molecular weight hemicelluloses, starches, and sugars. The lumen also absorbs soil, particularly fluids by capillarity [4].

A typical cotton consists of the following:

Cellulose	88–96 [1]
Protein	1–2
Organic acids	0.5–1
Sugars, pectins, wax	1.5–2.5

[1] Compositions indicated in weight (%).

The cellulosic membrane may oxidize at elevated temperature, producing color bodies. Another possibility is pyrolysis, essentially a dehydration of the cellulose molecules, producing color and fumes of acrolein or similar degradation products. The carmelization of sugar is a simple example of a pyrolytic process. Scouring products themselves do not initiate such reactions, but the combination of scouring process conditions of temperature, air, and the cleaning products may [5,6].

8.2.1   Scouring

Cotton fiber must be scoured in the raw to remove vegetable oils and incidental dirt. This is done by use of kiers, which are large pressure vessels that can be operated at 2–3 atm. The cotton is introduced with water and a scouring agent, then heated to 125–130 °C for about 6 hours. The processing time is extended to 8–12 hours for open kiers, which operate at atmospheric pressure and therefore cannot reach a temperature above 100 °C [4,5,7]. A typical scouring agent formulation is:

Sodium hydroxide	2.0 [1]
Sodium metasilicate	0.5
Wetting agent	0.5–1.0

[1] Compositions indicated in weight (%).

Wetting agents commonly used have hydrophile–lipophile balances (HLB) in the range of 12–14; linear alkylbenzenesulfonate (LAS) and alkylphenol ethoxylates are examples. The fats in the fiber provide additional soap as a result of saponification by the alkalinity of the scouring medium.

Air must be excluded during scouring because the oxidation of cellulose would degrade the cotton, resulting in the generation of oxycellulose and color development. Air can also result in the depolymerization of cellulose to hemicellulose.

Cotton goods are often scoured in J boxes (Fig. 8.1). In this method, the scour is used at about twice the concentration stated above, often padded on the fabric prior to loading into the J box. The goods are steamed for about an hour, rinsed, cooled, and then bleached using hydrygen peroxide. Hydrogen peroxide bleaching is commonly carried out with NaOH and silicate (see Chapter 4, Bleach) [8,9]. To save time, continuous bleaching and scouring can be carried out simultaneously in J boxes (Fig. 8.2). When the two steps are combined, however, use of bleach is increased to the extent that the soil present can react with it.

There is a trend in scouring toward using foam methods. Here the woven or knitted goods are treated continuously with a froth containing the scour. In addition to NaOH and metasilicate, a high-foaming surfactant such as alcohol sulfate or amphoteric is added to provide a stable, dense foam. This method should be used only for lightly soiled cloth because the temperature and time are not sufficient to clean heavily soiled fabric.

8.2.2   Size Removal and Scouring

Sizes used in processing cotton include starch and poly(vinyl alcohol). Removal of poly(vinyl alcohol) with simultaneous bleaching can be accomplished using the following active ingredients [7,9]:

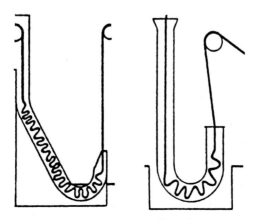

Figure 8.1   J-box configurations.

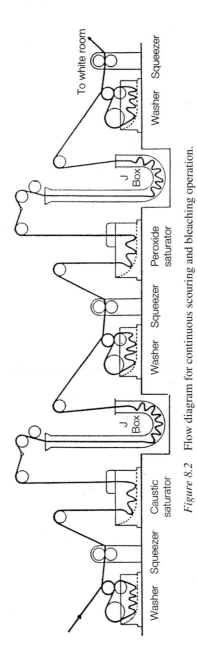

Figure 8.2   Flow diagram for continuous scouring and bleaching operation.

Sodium peroxide	4.0 [1]
Hydrogen peroxide	0.2
Sodium silicate	0.2
Wetting agent	0.5

[1] Compositions indicated in weight (%).

The wetting agent commonly used is nonylphenol ethoxylate, having an HLB of 12–14.

Starch sizes may be removed by boiling water, again with the addition of wetting agents, silicate, and peroxide as above.

An alternative, and increasingly popular method, is the use of enzymes to remove starch. Amylose is used, generally at 20–30 °C, and the pH is adjusted carefully to prevent the reduction of enzyme activity. Here the formulator must follow the recommendations of the enzyme supplier carefully for permissible temperature and pH ranges; otherwise enzyme activity will be lost. Since a number of different organisms can be used to generate the enzymes available, more specific directions cannot be given here. Enzyme treatment should generally be considered a separate step, to be followed by scouring.

8.3   Polyacrylics

Polyacrylic fibers are made by the polymerization of acrylonitrile; the process constitutes a homopolymerization except for the addition of low concentrations of comonomers, up to 15% by weight, for ease of dyeing. Redox catalyst systems are generally used for the polymerization. The comonomers used belong to many functional classes, having in common the presence of a vinyl group. Thus, substituted acrylic acid, acrylamides, and positively charged species such as vinyl pyridines or amines will allow dyes to react with the surface, a property generally referred to as dye substantivity.

Acrylic polymer cannot be spun from the melt because it decomposes near its melting point. Wet spinning is used, in which the polymer is dissolved in a polar solvent such as dimethylformamide or dimethylacetamide. The solution is forced through a spinneret into a bath of nonsolvent. Any number of liquids can be used, including water, glycols, alcohols, or even hydrocarbons. The swollen fiber is then dried.

The fiber may also be formed through dry spinning. Here the dissolved polymer is spun at elevated temperature to evaporate the solvent, which is recovered. Excess solvent remaining on the fiber is then removed by a hot water wash.

The dyes used depend on the comonomer charge. Since the comonomer may be anionic, cationic, or nonionic, there is a wide variety of dye systems available. Fiber finishes include softeners and antistats. Acrylic fibers are used alone or, frequently, blended with wool. Unlike wool, acrylic fibers do not shrink and can be dry-cleaned readily or laundered. Since the polymer has a low glass transition temperature, process temperatures must be kept below 80 °C.

8.3.1   Modacrylics

Modacrylics are copolymers of acrylonitrile and vinyl chloride. Emulsion polymerization yields a powder, which is dissolved in acetone and extruded into a water bath. The fiber is stretched to increase crystallinity and, therefore, strength. It is then annealed in heated air to set its structure [10,11]. Dyes used are cationic, disperse, or metallized, with the result that a wide range of colors is available. Modacrylic is commonly used as a fur substitute, as a furlike coat liner, and even for inexpensive wigs.

These fibers are not sensitive to acids and bases but are sensitive to highly polar solvents such as acetone. Modacrylic is quite resistant to oxidizing agents.

8.3.2   Scours

Cleaning of modacrylic fibers by severe scouring is seldom needed. As is the case with most synthetics, soil is not present unless the goods have been allowed to gather dirt from the surroundings. Any appropriate cleaning formulation may be used, bearing in mind that polar solvents such as acetone must be excluded. For example, to remove incidental soil, a product such as the following should prove adequate:

Sodium carbonate	5.0 [1]
Sodium metasilicate	2.0
Sodium tripolyphosphate	2.5
Wetting agent	1.5

[1] Compositions indicated in weight (%).

The wetting agent could be the LAS type, if foam is desired, or a nonionic (HLB about 13) for lower foam levels. If no foam is wanted, the formulation can be augmented with a low HLB nonionic surfactant such as a linear alcohol with two to three ethylene oxide groups [12,13].

8.4   Polyamides

The polyamide class of fibers is typified by the polymer resulting from the condensation of hexamethylenediamine and adipic acid (nylon 6-6) or the self-polymerization of caprolactam. The carbon chain lengths can be varied, depending on the mechanical and hand-feel properties desired. The inclusion of aromatic diamines as monomers gives results in the heat-resistant, exceptionally strong Aramid type of fibers. Nylon 6-6 is the type most commonly used for consumer textile applications.

Nylon polymers will dissolve in polar solvents to a greater or lesser degree, depending on the polymer type. Such solvents include alcohols, ketones, ethers, and chlorinated hydrocarbons. At elevated temperatures, both strong acids and bases will cause degradation. The combination of elevated temperature and light will also degrade the polymer, resulting in yellowing. Additives are used, mixed into the molten resin, to retard photochemical yellowing.

The fibers are extruded with many different cross sections, depending on use. The yarn denier can be as fine as silk or rather coarse. All these variations are dictated by the need for strength, feel, and the type of fabric to be made, whether it be for carpet or an evening gown.

Many of the uses for polyamide fibers derive from its strength, which is the result of its high degree of crystallinity after drawing near its glass transition temperature. Nylon blends with natural fibers are less common than polyester blends, but they are used.

The dyeability if nylon derives from the highly polar surface of this polymer. The phthalocyanines are examples of organic dyes used successfully. More commonly, pigments are included in the molten resin prior to spinning, since this method assures color fastness and also protects the fiber against light degradation. Also included in the melt are light stabilizers and opacifiers, as needed [10,11].

Nylon, being a highly crystalline fiber, must not be heated to temperatures close to its glass transition temperature. Such heating will reduce crystallinity, resulting in loss of strength.

8.4.1   Scours

Polyamide fibers, like most synthetic fibers, do not require severe cleaning. Reasonably high temperatures can be used. For example, nylon 6-6 will retain its strength for about 100 hours at 93 °C. Since, however, there are grades of polyamide having glass transition temperatures as low as 50 °C, extreme temperatures and residence times should be avoided. In the polymeric form, the amide linkage is reasonably stable toward acids and bases. But here too, extremes should be avoided. For general purposes, an adequate scour to remove incidental soil can be formulated somewhat as follows:

Sodium tripolyphosphate	3.0 [1]
Sodium metasilicate, anhydrous	2.0
Wetting agent	1.5

[1] Compositions indicated in weight (%).

The wetting agent could be LAS or nonionic surfactant with an HLB of about 13 [13,14].

8.5   Polyesters

Polyester is the result of the esterification of terephthalic acid with lactones or diols, commonly ethylene glycol. In practice, dimethyl terephthalate is used; thus the reaction is an ester interchange. Once the methyl alcohol has been recovered, the molten polymer may be extruded to form fibers. Additives to the molten resin that give opacity, reduce gloss, and so on, need not be discussed in detail here. The fiber is usually supplied by the manufacturer in either staple or monofilament form. Partially oriented yarn (POY) has been drawn to introduce a low degree of crystallinity, and therefore strength. Further drawing, texturing, and twisting then takes place, generally in another location. Polyester fibers are highly crystalline after spinning and drawing near the glass transition temperature. This crystallinity results in greater strength, which also allows for lower cross-sectional diameter, or finer denier.

The surface of polyester fiber is hydrophobic. Hence, the usual dyeing methods that involve water-soluble dyes cannot be used. Rather, we use "disperse dyes," which are actually fine pigment particles that become embedded in the surface of the fiber. This method of dyeing is important to the scouring step because of the stability of the finished product. The degree to which the dye is embedded in the fiber depends on the dyeing conditions and also on the crystalline nature of the fiber. Variations in fiber crystallinity result in "barre" effects, particularly noticeable in knitted goods as light streaks in the fabric, which appear after laundering has removed the pigment particles that were not attached firmly to the fiber [10,11]. Improper "overdying" of fabric can also lead to color removal during scouring, since the added dye is not evenly attached to the fibers.

When pigment has been added to the molten resin, barre effects are highly unlikely, giving the formulator more latitude in choice of scour.

The crystallinity of polyester is important also in terms of further treatment. Any heating of the goods to temperatures near the glass transition point will reduce crystallinity, degrading the mechanical properties of the fiber [7].

As a result of its manufacture, polyester, like other synthetic fibers, does not carry soil in the same sense as the natural fibers. The hydrophobic nature of its surface resists normal soiling by polar materials, but oils will stain the fiber, even entering the fiber structure, given enough time or heat to penetrate. The removal of oily stains is an important factor in the laundering of garments containing polyester (see Chapter 6).

Polyester blends well with cotton, common compositions being 40–60% polyester. Blends of polyester with wool are also marketed to a lesser extent. The strength of polyester is the main encentive for such blends. The natural fiber provides hand-feel and comfort and will absorb moisture, whereas polyester does not. The blends, of course, are formed using staple yarn, not monofilament.

8.5.1   Scours

Polyester, like other synthetic fibers, does not require severe scouring. Its hydrophobic nature results in excellent resistance to aqueous formulations, be they acidic or basic. The prevalence of polyester/cotton blends does present a soiling problem, as does the tendency of polyester to absorb oils.

To remove soils from polyester/cotton blends, common cotton scours may be used (see Section 8.2.1). For the removal of oily stains, an alkaline formulation with a high HLB surfactant such as LAS should be the first approach [13]. If this fails, an alcohol treatment may suffice. Oily stains tend to persist, however, especially if the oil has been in contact with the polyester fiber for some time.

For polyester/wool blends, similarly, the scour should follow the restrictions applied to wool (see Section 8.8), where high alkalinity is avoided, relying on wetting and polyphosphate or silicate as builder.

8.6   Polyolefins

Polyolefins include polyethylene and polypropylene, though the latter is of most interest in textile applications. Polypropylene is made by catalytic polymerization, using the Ziegler–Natta process or some modification thereof. Stereospecific polymers are formed with definite crystalline character, termed atactic, isotactic, or syndiotactic. Depending on crystal structure, a range of physical fiber properties is available, including strength, hardness, and resilience [15,16].

The fibers are used for carpeting and other applications requiring good wear characteristics, water resistance, and soil resistance. Polypropylene is hydrophobic and resistant to polar solvents. Low molecular weight esters, petroleum fluids, and chlorinated hydrocarbons will attack the fibers, initially swelling them by absorption, then gradually dissolving them.

Polypropylene fibers are formed by extrusion from the molten resin. Pigments and opacifiers are added to the melt to produce colored fiber. The thermoplasticity of the polymer allows shape and texture variations by varying fiber cross-sectional shape and diameter.

8.6.1   Scours

As with other synthetic fibers, soil is absent from polyolefins except for lubricants and incidental dirt. For soil removal, high pH may be used if required by the soil. Generally, however, a simple formulation with a nonionic surfactant (HLB 13 or higher) built with a polyphosphate such as sodium tripolyphosphate or pyrophosphate will suffice.

8.7   Rayons

Rayon, derived from cellulose, is the oldest of the synthetic fibers. In its manufacture, cotton, or more frequently, bleached sulfite wood pulp, is reacted with sodium hydroxide to "mercerize" it, forming alkali cellulose. The fibers are swollen at this point. Further reaction with carbon disulfide yields cellulose xanthate, soluble in alkaline solution. Extrusion of this "viscose" solution into an acid bath results in fibers of cellulose, as fine in structure as the spinneret used will permit.

Many modifications of the foregoing process chemistry exist to arrive at higher strength fibers, finer fiber structures, and other properties needed. Probably the best known rayon is acetate rayon, made from cellulose acetate [15].

8.7.1   Acetate

Acetate fiber is made by the reaction of cellulose with acetic anhydride, giving a product having one to three acetyl groups per glucose unit. Sulfuric acid catalyzes this reaction. The addition of excess water halts the reaction and precipitates the cellulose acetate. The resulting flakes are then dissolved in acetone or methyl acetate and "dry-extruded." This means that the fiber forms as the solvent evaporates. As with rayon, cross section and shape can be controlled by spinneret design. Drawing can increase strength, and the fibers may be given permanent twist.

Acetate fibers are dense, having no lumen. The extrusion spinnerets can give the fibers many different cross sections, and the fibers can be drawn to increase crystallinity and thus increase strength. Additives such as titania are mixed into the viscose solution to give opacity and, similarly, pigments may be added to spun-dyed fiber [15].

Rayon may be dyed with direct dyes similar to those used with cotton. Acetate is most often dyed with disperse dyes. Printing is very popular with both rayon and acetate.

8.7.2   Scours

As implied above, in scour formulations polar solvents such as acetone and esters must be avoided because the fibers are soluble. Strong acids and bases will also degrade the fibers. The triacetate is the more stable toward alkalinity, tolerating pH values up to 9.5 at room temperature. Dry cleaning may degrade the fiber if trichloroethylene is used.

It must also be remembered that printing is widely applied to rayon goods. Therefore, bleaches should not be used. A mild scour for such fibers is simply to use a surfactant with an HLB above 13. For undyed and unprinted goods, a low degree of alkalinity can be tolerated, which can be achieved by using polyphosphates, silicates, or sodium carbonate.

8.8   Wool

Wool is a fiber derived from animal pelts. Commonly it is from sheep or goat hair, although camel, llama, vicuña, horse, and other pelts are used, depending on the part of the world and the intended use of the finished material. Goat hair, prized for its fiber length and softness, is used in luxury fabrics. Horse hair, being strong and stiff, is used for reinforcement, as in the inner liner of men's suiting. Soft, pliable camel hair is used for expensive suits and coats; it has superior insulating properties. Blends of these fibers are common, to take advantage of the best features of each.

Wool has a structure different from the fibers discussed earlier. The internal structure is complex, with various longitudinal layes of proteinaceous components, the important ones being the inner cortex (the source of structure and strength) and the outer cuticle (a scaly material, which gives the fiber its directionality). Directionality is expressed in the fact that wool's friction is far greater going from the tip of the fiber toward its root than in the opposite direction. This property aids wool in its ability to retain yarn structure so that it may be loosely or tightly spun. While aiding with respect to directionality, the cuticle is also responsible for wool's tendency to shrink. The mechanical flexing of wool during washing causes the individual fibers to move with respect to one another, and the scales interlace strongly when fibers lying in opposite directions rub against each other, compacting and thus shrinking the structure of the fabric. Shrinkage of the individual fibers is negligible; it is the fabric structure that is affected.

In scouring, the treatment of raw wool is therefore quite different from that of wool fabric. Changes in relative humidity and temperature will cause the individual fibers to contort, curling or straightening. Fiber structure may be set by crosslinking with thiols, as in treatments for crease resistance. Wool is the most complex of fibers from a chemical standpoint. It consists of protein molecules with disulfide linkages bridging the protein molecules to increase molecular weight. Cleaning methods must take these factors into account. The disulfide bonds are sensitive toward degradation by alkali, resulting in a depolymerized structure lacking the necessary mechanical properties [17].

The scouring of wool results in the recovery of a grease that has considerable commercial value. It is high in fatty acid esters and primarily saturated, melting at 35–40 °C, which is close to the normal temperature of the human body. Raw grease is used commercially (as Degras) as a lubricant in steel rolling, for example, and as a source of fatty acids and alcohols. A refined grade of wool fats, lanolin, sees application in cosmetics and as a constituent of medical salves [18].

Raw wool is badly soiled, thanks to its origin. Animal fats and oils are present, the ether-soluble fraction being called wool-wax and the water-soluble, squint. Additionally there is dirt from the pasture. A typical sheep wool is composed of:

Oven-dried wool	65–70 [1]
Wool-wax	20–25
Squint	2–14
Moisture	10–15

[1] Compositions indicated in weight (%).

8.8.1   Scours

The removal of the wax, squint, and dirt must take into account the preservation of the fiber properties. Wool, being proteinaceous, is attacked by strong alkali, softening and swelling, with reduction in strength. Therefore, only weak alkalinity may be used [5,17]. A typical first scour is composed of:

Wetting agent	0.8 [1]
Sodium carbonate	0.2

[1] Compositions indicated in weight (%).

The scouring temperature is usually limited to 50 °C to prevent damage to the wool fiber. The wetting agent is commonly soap, LAS, or an alkylphenol ethoxylate with an HLB of about 13.

The first scour is followed by additional scouring steps in which the surfactant concentration is lowered and that of the sodium carbonate is maintained. Rinsing is with warm water. Ammonia may be added if additional mild alkalinity is needed.

For scouring of finished goods such as knits, warm water with 0.3–0.6% surfactant is used, with ammonia added if needed. Other weak bases such as sodium silicate and polyphosphates can be considered. Again, the bath pH is kept below 10.5 to prevent degradation. Silicates of about 2.0 SiO_2/Na_2O or sodium tripolyphosphate are suitable as the alkaline component. Ammonia should not be added to a silicate-containing wash because it may cause a powdery silica precipitate [19].

8.8.2   Proteolytic Enzymes

Proteolytic enzymes, used for wool scouring, decompose the residual protein fragments and lower molecular weigh proteins that may be present on the fibers. The literature contains references to their use in normal scouring operations as additives to the mildly alkaline scours mentioned above. Care must be practiced in their use, however, to avoid damage to the fiber by digestion of the proteins making up the fiber structure itself.

1. Too high an enzyme concentration and long contact times may result in attack on the cuticle or even the fiber.
2. The pH, the temperature, and the presence of other chemicals have strong influence on enzyme activity.
3. Enzymes derive from many bacteriological and plant sources. What is correct for one type of enzyme may be incorrect for another. The formulator needs to be aware of the specifics of the enzyme to be used, with respect to the proper conditions of temperature and pH. The supplier's literature is the best source for such information.

It is difficult, therefore, to give specific formulation information for enzymes. In addition, the formulator must be clear on the purpose of the treatment. The inclusion of enzyme in the initial scouring of raw wool, aftertreatment as a separate scour, treatment of the wool after spinning, and treatment after weaving or knitting are all alternatives. The point at which the enzyme treatment is to be applied will influence the formulation because of the nature and amount of soil present and because of the conditions of application. Proteolytic enzymes can be very useful if applied properly, but attention to detail is needed to maintain enzyme activity and to achieve the desired result.

8.9   The Role of Cationic Surfactants

This chapter has been concerned with scours, and cationic surfactants have not been mentioned, with good reason. They do not see application in scours. However, there is an area for the application of cationics, and that is in finishing—the final treatment a fabric receives to impart customer appeal, including softness, drape, or antistatic properties [2,18]. See Chapter 3, Section 3.5 (Cationic Surfactants) for more details.

Antistatic properties, the ability to dissipate static electric charge, are almost always achieved with cationics. These include quaternary compounds and amine salts. The quats used are either long-chain tertiary amines or imidazolines that have been quaternized. Amines may be either primary or secondary, reacted with hydrochloric or acetic acid, commonly. While they are particularly effective on cotton, being substantive thanks to charge neutralization, the long hydrocarbon chains are attracted to polyester and other synthetic fabrics. They will adsorb on wool but will be removed by laundering.

Long-chain quats and quaternized imidazolines are used as softeners for goods of all kinds, though the foregoing observations regarding different fabrics apply. Both "hand" and drape are improved, lending the fabrics customer appeal.

Cationic surfactants are generally sold as emulsions, since they are not water soluble. Hence emulsifying agents, such as nonionic surfactants, are present. It is probable that mixed micelles are formed which adsorb, rather than the active salts or quats alone. For antistatic effectiveness, much of the fiber surface must be covered, to permit hemimicelle formation to function as the effective mechanism. For softness, the adsorption should take place at points where fibers come into contact with each other. This implies colloidal particles at junction points. In practice, probably both types of adsorption occur.

References

1. Schick, M.J., Ed., *Surface Characteristics of Fibers and Textiles,* Dekker, New York, 1977.
2. Tortora, P.G., *Understanding Textiles,* MacMillan, New York, 1987.
3. Cook, J.G., *Handbook of Textile Fibers,* 5th ed., Merrow Publishers, Durham, U.K., 1984.
4. Hanby, D.S., *American Cotton Handbook,* Wiley-Interscience, New York, 1965.
5. Trotman, E.R., *Dyeing and Chemical Technology of Textile Fibers,* 6th ed., Wiley, New York, 1984.
6. Jakobi, G., Lohr, A., *Detergents and Textile Washing,* VCH Verlag, Weinheim, Germany, 1987 (Trans. Russey, W.E.)
7. Prager, W., Blom, M.J., Preparing cotton and polyester/cotton for printing, Text. Chem. Color., *11*, 28 (1979).
8. Easton, B.K., Preparation and bleaching, Text. Chem. Color., *13*, 15 (1981).
9. Rowe, M.H., Desizing/scouring with hydrogen peroxide. Text. Chem. Color., *10*, 215 (1978).
10. Lyle, D.S., *Modern Textiles,* Wiley, New York, 1976.
11. *Encyclopedia of Textiles,* Prentice-Hall, Englewood Cliffs, NJ, 1980.
12. Leuk, J.F., Sequestrants in dyeing and finishing, Am. Dyestuff Rep., *69*, 49 (1979).
13. Sabia, A.J., Non-ionic surfactants in textile processing, Text. Chem. Color., *12*, 22 (1980).
14. Kowalski, X., Sequestering agents in bleaching and scouring, Text. Chem. Color., *10*, 161 (1978).
15. *Kirk-Othmer Encyclopedia of Chemical Technology,* Wiley-Interscience, New York.
16. Ahmed, M., *Polypropylene Fibers–Science & Technology,* Elsevier, New York, 1982.
17. von Bergen, W., *Von Bergen's Wool Handbood,* Wiley-Interscience, New York, 1984.
18. Rouette, H-K., Kittan, G., *Wool Fabric Finishing,* Engl. transl., Wool Development Institution, Ilkley, U.K., 1991.
19. Lewis, D.M., Ed., *Wool Dyeing,* Society of Dyers & Colourists, Bradford, U.K., 1992.

CHAPTER 9

The Detergent Regulatory and Environmental Situation

Edwin A. Matzner

This chapter presents an overview of detergent and cleaning product regulation in the United States and Canada. Only enacted and significant legislation applying to ordinary consumer detergent products is covered. Pending (unenacted) legislation is briefly presented for the sake of completeness but is of no significance, since its outcome is uncertain.

No legal advice can be given by simplfying statutes into layman's terms. Especially in the area of phosphate legislation, every single law differs from every other one in what is banned, how exemptions and residual levels are described, and so on. As a result, this chapter quotes the actual language of regulations in every case, to permit individual evaluation of exactly what a ban covers. No guarantee of completeness of the listings is possible in this still-developing area.

9.1   Introduction

The history of detergent legislation is largely that of phosphate regulation. The initial environmental problem, surfacing in the 1960s, was that in certain areas, phosphate wastes generated by homes and businesses contributed to algal growth, causing eutrophication. The difficulty is that every living organism contains, ingests, and excretes large amounts of phosphate: moreover, the environment requires continuous addition of even large amounts of phosphate for farming and food production. On a national basis, detergents contribute less than 5% of the phosphate in the environment. In areas where phosphate removal is part of sewage treatment, the contribution of detergents is even smaller. Adequate sewage treatment is the most satisfactory solution to the problem mentioned, but what in fact developed was regulation of consumer products.

Detailed regulation of consumer products is generally the result of public opinion in a given area. Public opinion, in turn, is predominantly derived from media treatment of that area. In presenting problems, the media simplify complex technical situations to make them clear to a lay public. They then suggest solutions, which may generate public pressures favoring these solutions.

Cleaning agents, most of which are eventually discharged into water, are very much in the public eye. Cleaning agents are consumer products familiar to most people. Water is likewise a universal entity. Early media definition soft-pedaled product benefits and downplayed the fact that the products attacked were harmful to insects, pests, dirt, bacteria,

Edwin A. Matzner, P.O. Box 3230, St. Louis, Missouri 63130, U.S.A.

or weeds. Instead, the products were presented as harmful to children and animals, with ill effects on pure drinking water, the planet, and so on. Then worried and concerned citizens felt obliged to take action.

The basic trend in regulation, then and today, is to define as "hazardous" or "toxic" or "corrosive" some material that, by factual criteria, does not necessarily have these attributes. Early descriptions of phosphates in detergents thus resorted to presenting phosphate, an essential nutrient, as a toxic, deadly substance. Naturally, such publicity generated unfavorable opinion and public pressures, which are well established today, against phosphates and other detergent ingredients.

A substantial network of regulation exists at both the federal and state levels to ensure safety and environmental acceptability of consumer products. Once material such as phosphate has been defined as a poison by laymen, this network of opinion and regulation springs into action. No matter that no life can exist without phosphates (DNA, the genetic material contained in every single cell of every organism, typically contains 28% phosphate) or that 40 tons of phosphate had to be added to Lake Mead to permit fish life [1]. The public now sees phosphates as toxic and hazardous substances to be banned.

Nutrient problems in water pollution are a critical issue in water management and require advanced sewage treatment, which includes phosphate removal and water quality standards [2] to keep pace with the urban and industrial development that cause the problem in the first place. Regulating the contribution made by detergent phosphate is but a tiny step toward this goal.

9.2   United States

9.2.1   General Regulations

Most of the federal regulation of consumer products falls under the following major laws:

- The Federal Insecticide, Fungicide, and Rodenticide Act (FIFRA), administered by the Environmental Protection Agency (EPA). Disinfectant and antimicrobial cleaning products are now classified as pesticides under FIFRA.
- The Federal Food, Drug, and Cosmetic Act, administered by the Food and Drug Administration (FDA).
- The Federal Hazardous Substances Act and the Poison Prevention Packaging Act, administered by the Consumer Product Safety Commission (CPSC).
- The Fair Packaging and Labeling Act, administered by the Federal Trade Commission (FTC).
- The Hazardous Materials Transportation Act, adminstered by the U.S. Department of Transportation (HAZMAT).

Besides the obvious aims of protecting consumers and the environment, no provision of these acts presently regulates the production and distribution of detergent and cleaning agents as directly as the other legislation described in this chapter. Nevertheless, the possible relevance of these laws to any particular product should be verified.

Other federal legislation that should be checked for possible relevance includes the Clean Air Act, the Water Pollution Control Act (Clean Water Act), the Marine Sanctuaries Act, the National Environmental Policy Act, the Noise Control Act, the Occupational Safety and

Health Act, the Resource Conservation and Recovery Act (RCRA) or Solid Waste Disposal Act, the Used Oil Recycling Act, the Safe Drinking Water Act, the Superfund or Comprehensive Environmental Response, Compensation, and Liability Act (CERCLA), the Superfund Amendments and Reauthorization Act (SARA), and the Toxic Substances Control Act.

Some cleaning agents use solvents that require checking against the Clean Air Act solvent/ozone amendments, as well as the National Protocol on Substances That Deplete the Ozone Layer.

9.2.2   FTC Environmental Claim Guidelines

The Federal Trade Commission, in issuing the Environmental Marketing Guidelines, seeks to protect consumers, to bolster their confidence in environmental claims, and to reduce manufacturers' uncertainty about which claims might lead to FTC enforcement actions. Working in concert with the EPA and the White House Office of Consumer Affairs, the FTC has published [3] voluntary environmental labeling guidelines, addressing the terms as "degradable, biodegradable, photodegradable, compostable, recyclable, recycled content, source reduction, refillable, ozone safe, ozone friendly, etc."

A novel principle in arriving at the guidelines was the use of consumer polls to determine how the public understood the meaning of such environmental words. The definition of a word was then based on this understanding, and deviations from these consumer-generated definitions were considered to be deceptive. In addition, any term used without qualification is thought by the Guidelines to connote total applicability.

- *Example 1:* If a manufacturer increases the recycled content of a package from 2 to 3% recycled material, the claim "50% more recycled content than before" is deceptive because 50%, although technically correct, conveys an impression of significance.
- *Example 2:* An unqualified claim of "recyclable" is deceptive even if collection sites are established in a significant percentage of communities or available to a significant percentage of the population. Collection sites must be available to a substantial majority of consumers and communities.
- *Example 3:* The claim of "degradable" on a trash bag is deceptive if the marketer relies on soil burial tests that show the product will decompose in the presence of water and oxygen, even though trash bags are customarily disposed of in incineration facilities or sanitary landfills.
- *Example 4:* The unqualified claim "contains no CFCs" is deceptive if the product indeed contains no chlorofluorocarbons, yet does contain HCFC-22, another ozone-depleting ingredient.

9.2.3   Surfactant Regulation

Section 313 of the federal Emergency Planning and Community Right-to-Know Act (EPCRA) determines the leading "toxic" chemicals in the United States. While intended as a reporting requirement, the list of toxic materials produced under Section 313 is generally looked on as one of the leading compendia of hazardous chemicals, and a growing body of state legislation mandates the replacement of chemicals so declared as "toxic."

The present definition of toxic glycol ethers under Section 313 includes some nonionic surfactants (alkyl ethers, α-olefinsulfonates, alkylphenol ethoxylates). As the result of the submission of safety data, EPA was reported to have decided in spring 1992 to revise the glycol ether definition to exclude surfactants. This decision had not been published at this writing and may not include revision of the definition under other statutes, such as Superfund and the Clean Air Act, that have followed the definition of "toxic" in Section 313.

9.2.4   Boron Regulation

Boron is an ingredient of detergents that has been found to be a reproductive toxicant at extremely high dosages. Boron also exhibits phytotoxicity and toxicity to fish. EPA has issued a health advisory on boron and proposed it as one of the 25 chemicals for which a Safe Drinking Water Act standard must be issued by mid-1993. EPA's present drinking water guideline is 0.9 ppm, which may be lowered. The World Health Organization is proposing an 0.3 ppm boron advisory by the same time. Direct treatment of drinking water or sewage effluent to remove boron is feasible but likely to be very costly. Source control by product bans or ingredient limitations is a possible future option.

9.3   U.S. State and Local Regulations

9.3.1   State and Local Phosphate Regulations: Enacted

To give the reader an overview, Fig. 9.1 shows in graphical form the history of detergent phosphate ban enactment from 1971 to 1993. The population living in ban areas is shown as a percentage of 1985 population values. Only major population areas are shown. Tiny municipalities are omitted. Seventeen major areas have enacted bans, identified in timing and location in Fig. 9.1. The first of these in time, Dade County, Florida, was also the first to repeal its ban. It is thus shown as a 1985 negative in the cumulation. Between 1971 and 1993, there have been 68 rejections of bans by major areas. Both the timing and the location of these occurrences are shown.

Sections 9.3.1.1–9.3.1.25 present details of the important phosphate bans enacted in the United States. The states or other jurisdictional areas are listed in alphabetical order. It is important to note that except for Austin and LaGrange, in Texas, no two bans are identical in their provisions. Even definitions of specific exemption areas (health care, metal treatment, food and dairy, etc.) differ in their exact scope from ban to ban. Some bans permit no phosphorus in detergents; others allow up to 0.5% phosphorus, or up to 2% phosphate. Enforcement of provisions and of entire bans is also quite variable. While opinions can be rendered, no legally reliable summaries are possible. The actual wording, given below, permits a partial interpretation of the ban status in a given area.

Another factor to consider when reviewing ban language is the confusion between chemical terms. The reader will find "phosphorus," "phosphorous," "phosphorus pentoxide P_2O_5," and "phosphates," with some language referring to quantities of these materials. Here are useful approximate conversion factors for quantities: 1 lb of elemental phosphor*us* (chemical symbol "P") is equivalent to about 4 lb of a typical sodium phosph*ate*, or about 3 lb

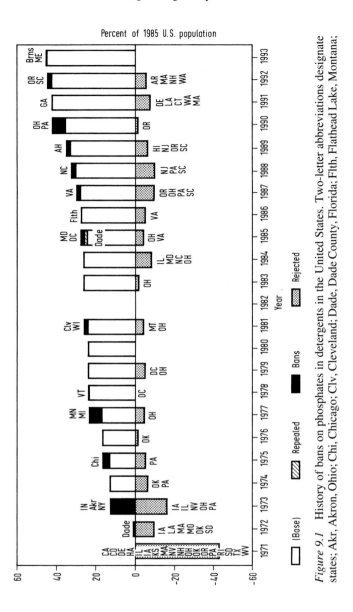

Figure 9.1 History of bans on phosphates in detergents in the United States. Two-letter abbreviations designate states; Akr, Akron, Ohio; Chi, Chicago; Clv, Cleveland; Dade, Dade County, Florida; Flth, Flathead Lake, Montana;

of 100% phosphoric acid, or to 2.3 lb of P_2O_5. Although *phosphates contain phosphorus*, their chemical, nutrient, and toxicological properties have absolutely nothing in common with those of elemental phosphorus. The term "elemental phosphorus", which occurs in many bans, is used only quantitatively as a basis for numbers. Note that these bans often wrongly substitute "phosphorous" for "phosphorus;" the former term is an adjective describing compounds of trivalent phosphorus, which are never employed in detergents.

Population figures given below are generally those for 1991 [5], when the total U.S. population was 252 million.

9.3.1.1   Akron, Ohio (population 660,000)

Still in force, not replaced by Northern Ohio ban effective January 1990 (per Section 9.3.1.17).
Definition of cleaning agent (in which phosphates are banned):

> Any cleaning compound which is available for household use, laundry use, other personal uses or industrial use, which is composed of organic and inorganic compounds, including soaps, water softeners, surface active agents, dispersing agents, foaming agents, buffering agents, builders, fillers, dyes, enzymes and fabric softeners, whether in the form of crystals, powders, flakes, bars, liquids, sprays or any other form.

Periodically renewed variances are granted by the Department of Public Health for detergents for use in:

- Machine dishwashers
- Dairy equipment
- Beverage equipment
- Food processing equipment
- Industrial cleaning equipment

9.3.1.2   Austin, Texas (population 380,000); LaGrange, Texas (population 3800)

Effective June 1991, Austin; effective August 1992, LaGrange.
Definition of cleaning agent (in which phosphorus is limited to 0.5%):

> Household laundry detergent, a laundering cleaning compound in liquid, bar, spray, tablet, flake, powder, or other form used for domestic clothes-cleaning purposes.

The term "household laundry detergent" shall not mean:

- A dishwashing compound, household cleaner, metal cleaner, degreasing compound, commercial cleaner, industrial cleaner, or other substance that is intended to be used for nonlaundry cleaning purposes;
- A detergent used in dairy, beverage, or food processing cleaning equipment
- A phosphorous [sic] acid product, including a sanitizer, brightener, acid cleaner, or metal conditioner
- A detergent used in hospitals, veterinary hospitals or clinics, or health care facilities or in agricultural production
- A detergent used by industry for metal cleaning or conditioning
- A detergent manufactured, stored, or distributed for use or sale outside of the state

- A detergent used in any laboratory, including a biological laboratory, research facility, chemical laboratory, and engineering laboratory
- A detergent used in a commercial laundry that provides laundry services for hospital, health care facility, or veterinary hospital

9.3.1.3   Branson, Missouri (population 2500)

Effective January 1993.

Definition of cleaning agent (in which elemental phosphorus is limited to 0.5%):
> Any cleaning agent (this means a laundry detergent, dishwashing compound, commercial cleaner, phosphate compound or other substance that is intended to be used for cleaning purposes) except those used in automatic dishwashers (which may contain up to 8.7% elemental phosphorus).

Exempted by one-year exemptions, requests for the continuation of which must be submitted to the Department of Health and Public Works by October 1 of each year:

- Detergents used in dairy, beverage or food processing cleaning equipment
- Phosphoric acid products, including sanitizers, brighteners, acid cleaners, or metal conditioners
- Detergents used in hospitals, veterinary hospitals or clinics or health care facilities, or in agricultural production
- Detergents used by industry for metal cleaning or conditioning
- Detergents manufactured, stored or distributed for use or sale outside the state
- Detergents used in any laboratory, including a biological laboratory, research facility, chemical laboratory or engineering laboratory
- Detergents used in a commercial laundry that provides laundry services for a hospital, health care facility or veterinary hospital

9.3.1.4   Chicago, Illinois (population 2,784,000)

Effective February 1971.

Definition of cleaning agent (in which phosphates are banned):
> Any agent available for household use, laundry use, other personal uses or industrial use, including soap, water softeners, surface active agents, dispersing agents, foaming agents, buffering agents, builders, fillers, dyes, enzymes, and fabric softeners.

Exempted by variance by the Department of Environmental Control, periodically renewed:

- Industrial detergents used in multiple purpose cleaners
- Institutional detergents used in health care facilities, such as sanitizers, disinfectants, delimers and on-site laundries
- Detergents for automatic dishwashing machines used in households, institutions, and commercial establishments

Exempted by variance by the Department of Environmental Control until state-of-the-art produces a substitute:

- Phosphoric acid products used in sanitizers, brighteners, cleaners and metal conditioners

- Clean-in-place (CIP) detergents used for the cleaning of process equipment in the food, beverage and dairy industries

9.3.1.5   District of Columbia (population 598,000)

Effective September 1986.
Definition of cleaning agent (in which phosphates are banned):
> Soaps and detergents used for domestic or commercial cleaning purposes, including the purposes of cleaning fabrics, dishes, eating and cooking utensils, homes, or commercial premises, but the term does not include cosmetics and personal hygiene products like toothpaste, shampoo, and hand soap.

Completely exempt from ban:

- Hospital or health care equipment laundry services
- Dairy, beverage, food processing equipment
- Other industrial and institutional applications

Elemental phosphorus limitation (8.7%):

- Dishwashing machines

Variances granted:

- Upon application to Director of Consumer and Regulatory Affairs if the application documents hardship, unreasonableness because an adequate substitute is not available, or disruption of scientific research.

9.3.1.6   Georgia (population 6,623,000)

Effective January 1991.
Definition of cleaning agent (in which phosphates are banned):
> Any cleaning agent.

Exemptions:

- Used in agricultural or dairy production
- Used to clean commercial food or beverage processing equipment or containers
- Used as industrial sanitizers, metal brighteners, or acid cleaners, including those containing phosphoric acid or trisodium phosphates
- Used in industrial processes for metal, fabric, or fiber cleaning and conditioning
- Used in hospitals, clinics, nursing homes, other health care facilities, or veterinary hospitals or clinics
- Used by a commercial laundry or textile rental service company or any other commercial entity
  a. to provide laundry service to hospitals, clinics, nursing homes, other health care facilities, or veterinary hospitals or clinics
  b. to clean textile products supplied to industrial or commercial users of the products on a rental basis, or
  c. to clean professional, industrial, or commercial work uniforms
- Used in the manufacture of health care or veterinary supplies

- Used in any medical, biological, chemical, engineering, or other such laboratory, including those associated with any academic or research facility
- Used as water softeners, antiscale agents, or corrosion inhibitors, where such use is in a closed system such as a boiler, air conditioner, cooling tower, or hot water heating system
- Used to clean hard surfaces including windows, sinks, counters, floors, ovens, food preparation surfaces, and plumbing fixtures
- Cleaning agents which are manufactured, stored, sold or distributed for uses other than household laundry detergents or household or commercial dishwashing agents
- Cleaning agents which contain elemental phosphorus in an amount not exceeding 0.5 wt % which is incidental to manufacturing
- Cleaning agents which contain elemental phosphorus in an amount not exceeding 8.7 wt % and which are intended for use in a commercial or household dishwashing machine
- Any natural or commercial fertilizers

9.3.1.7   Idaho Municipalities (estimated population 91,000)

Effective March–April 1989: Bonner County, Clark Fork, East Hope, Hope, Kootenai, Old Town, Ponderay, Priest River, Sandpoint; effective July 1990, McCall.
Definition of cleaning agent (in which phosphates are banned):
> Household cleaning product, defined as any domestic or commercial cleaning product, including but not limited to soaps, detergents, laundry bleaches and laundry additives used for domestic or commercial cleaning purposes, including but not limited to the cleaning of fabrics, dishes, food utensils and household and commercial premises. Household cleaning product does not mean foods, drugs, cosmetics or personal care items such as toothpaste, shampoo, or hand soap.

Exemptions:

- Automatic dishwasher detergent containing 8.7% or less elemental phosphorus
- Those products used for cleaning medical or surgical equipment or supplies in licensed acute care or long-term health care medical facilities

Coeur d'Alene, Post Falls (population 27,000), effective 1990.
Definition of cleaning agent (in which phosphates are banned):
> Any product including but not limited to soaps, detergents, laundry bleaches, laundry additives, and fabric softeners used for the purpose of cleaning laundry.

9.3.1.8   Indiana (population 5,610,000)

Effective February 1972.
Definition of cleaning agent (in which phosphates are banned):
> None.

Completely exempt from ban:

- Clean-in-place food processing and dairy equipment
- Phosphoric acid products including sanitizers, brighteners, acid cleaners, and metal conditioners

- Household and commerical dishwashing machine detergents
- Hospital and health care facility detergents
- Institutional laundry detergents
- Dairy, beverage, food processing, and other industrial cleaning equipment detergents
- Other uses of detergents in which the elemental phosphorus contents are not permitted to enter any public or private sewer or to be disposed of in the natural environment
- Detergents or cleaning compounds contained in fuel or lubricating oil

9.3.1.9   Maine (population 1,235,000)

Effective July 1993.
Definition of cleaning agent (which may not contain more than 0.5% elemental phosphorus by weight):
> Household laundry detergent: a cleansing agent used primarily in private residences for washing clothes.

9.3.1.10   Maryland (population 4,860,000)

Effective December 1985.
Definition of cleaning agent (in which phosphates are banned):
> A laundry detergent, dishwashing compound, household cleaner, metal cleaner, degreasing compound, commercial cleaner, industrial cleaner, phosphate compound, or other substance that is intended to be used for cleaning purposes.

Completely exempt from ban:

- Dairy, beverage, or food processing cleaning equipment
- Hospitals, veterinary hospitals or clinics, or health care facilities
- Agriculture production
- Phosphoric acid products, including sanitizers, brighteners, acid cleaners, or metal conditioners
- Industrial metal cleaning or conditioning laboratories
- Laboratories
- Commercial laundries providing laundry services for hospitals, health care facilities, or veterinary hospitals
- Detergents manufactured, stored, or distributed for use or sale outside of the state
- (Upon application, excluded by Secretary by rule or regulation) Substances which, if banned, would cause a significant hardship or be unreasonable because of the lack of an adequate substitute

Elemental phosphorus limitation (8.7%):

- Commercial and household dishwashing machine detergents

9.3.1.11   Michigan (population 9,368,000)

Effective July 1972.
Definition of cleaning agent (in which phosphates are banned):

A laundry detergent, dishwashing compound, household cleaner, metal cleaner, degreasing compound, commercial cleaner, industrial cleaner, phosphate compound, or other substances intended to be used for cleaning purposes.

Elemental phosphorus limitation (14%):

- Commercial automatic or machine dishwashers
- Dairy, agricultural, and farm operations
- Manufacture, preparation and processing of foods and food products

Elemental phosphorus limitation (28%):

- Metal cleaners, brighteners, and treatment compounds
- Conversion coating agents, corrosion removers, paint removers, rust inhibitors, etchants, phosphatizers, degreasing compounds, and industrial cleaners
- Commercial cleaners intended primarily for use in industrial and manufacturing processes

9.3.1.12   Minnesota (population 4,432,000)

Effective August 1979.
Definition of cleaning agent (in which elemental phosphorus is limited to 0.5%):
A laundry detergent, dishwashing compound, household cleaner, metal cleaner, degreasing compound, commercial cleaner, industrail cleaner, phosphate compound or other substance intended to be used for cleaning purposes.

Completely exempt from ban:

- Laundry detergents and other cleaning products intended for commercial, industrial, or institutional use

Elemental phosphorus limitation (11%):

- Detergents for automatic machine dishwashing

Elemental phosphorus limitation (20%):

- Chemical water conditioners: water-softening and anti-scale chemicals, corrosion inhibitors, or other substances intended to be used to treat water

9.3.1.13   Monroe County, Florida (estimated population 70,000)

Effective October 1991.
Definition of cleaning agent (which cannot contain more than 0–0.5% elemental phosphorus by weight):
Any detergent.

Completely exempt from ban:

- Any machine dishwashing detergent containing less than or equal to 1.1 grams of elemental phosphorus per tablespoon of detergent, determined by a Florida Department of Environmental Regulation method
- Industrial and institutional automatic dishwashing products exempted for an additional three months from effective date.

9.3.1.14   Montana

Effective December 1986 and November 1987, Flathead and Lake counties (population 70,000); effective May 1989, Missoula (population 80,000); effective January 1990, Superior (population 1000).

Definition of cleaning agent (in which phosphates are banned):

> Any product including but not limited to soaps and detergents, used for domestic or commercial cleaning purposes, including the cleaning of fabrics, dishes, food utensils, in household and commercial premises.

Exempt from ban:

- Beverage processing
- Health care services and facilities
- Commercial establishments (any trade, business, etc., including but not limited to laundries, hospitals, hotels, motels, and food or restaurant establishments)
- Cleaning medical or surgical equipment
- Dairy equipment
- Used in agricultural operations

Elemental phosphorus exemption (8.7%):

- Dishwashing detergents

Elemental phosphorus exemption (20%):

- Chemical water conditioners: water-softening material, or other intended to treat water for use in machines for washing laundry

9.3.1.15   New York State (population 18,058,000)

Effective January 1972.

Definition of cleaning agent (in which phosphates are banned):

> Any product, including but not limited to soaps and detergents, containing a surfactant as a wetting or dirt emulsifying agent and used primarily for domestic or commercial cleaning purposes, including but not limited to the cleansing of fabrics, dishes, food utensils and household and commercial premises.

Household cleansing products shall not mean:

- Foods, drugs and cosmetics, including personal care items such as toothpaste, shampoo and hand soap
- Products labeled, advertised, marketed and distributed for use primarily as pesticides.

Elemental phosphorus limitation (8.7%):

- Cleaning products used in dishwashers, food and beverage processing equipment, and dairy equipment

9.3.1.16   North Carolina (population 6,737,000)

Effective January 1988.

Definition of cleaning agent (in which phosphates are banned):
> A laundry detergent, dishwashing compound, household cleaner, metal cleaner or polish, industrial cleaner, or other substance that is used or intended for use for cleaning purposes.

Exempted:

- Cleaning agents used in agricultural or dairy production, commercial food or beverage processing equipment or containers, industrial sanitizers, metal brighteners, acid cleaners including those containing phosphoric acid or trisodium phosphate, industrial metal, fabric or fiber cleaning and conditioning
- Hospitals, clinics, nursing homes, other health care facilities, or veterinary hospitals or clinics, or funeral homes
- Commercial laundry or textile rental service companies providing laundry service to hospitals, clinics, nursing homes, other health care facilities, veterinary hospitals or clinics or funeral homes or to clean textile products supplied on a rental basis, or to clean work uniforms
- Manufacture of health care or veterinary supplies
- Any medical, biological, chemical, etc., laboratory
- As water softeners, antiscale agents or corrosion inhibitors in closed systems
- To clean hard surfaces, windows, sinks, counters, floors, ovens, food preparation surfaces and plumbing fixtures
- Commercial and household dishwashing products containing 8.7% elemental phosphorus
- Manufactured, stored, sold or distributed for use outside state

Exempted by Environmental Management Commission:

- Cleaning agents with 8.7% phosphorus upon finding that there is no substitute, that compliance would be unreasonable, or create a hardship.

9.3.1.17 *Northern Ohio (estimated population 6,000,000)*

Effective January 1, 1990: Ashtabula, Trumbull, Lake, Geauga, Portage, Stark, Cuyahoga, Summit, Medina, Lorain, Ashland, Richland, Huron, Erie, Crawford, Marion, Wyandot, Seneca, Sandusky, Ottawa, Lucas, Wood, Hancock, Hardin, Fulton, Henry, Putnam, Allen, Auglaize, Shelby, Mercer, Van Wert, Paulding, Defiance, and Williams counties.
Definition of cleaning agent (in which phosphates are banned):
> Household laundry detergent.

Exempted:

- Dishwashing detergents, rinsing aids, sanitizers
- Coating agents, corrosion or paint removers, rust inhibitors, metal brighteners, etchants, or other surface conditioners
- Solvent cleaner or other products not used with water
- Products used in institutions, hospitals, veterinary hospitals or clinics, nursing homes or other health care facilities or commercial laundries providing service for same
- Products used in food product manufacture, including dairy, beverage, egg, fish, poultry, meat, fruit, vegetable processing facilities
- Products for personal use, including soaps, dentrifices, shampoos, cleaning creams, toothpastes, etc.

- Products subject to registration or control under federal or state law governing foods, drugs, cosmetics, insecticides, fungicides, or rodenticides
- Window, oven, hard-surface cleaners
- Water softeners, antiscalants, corrosion inhibitors
- Products used as cleaners in the production of milk
- Products used for cleaning in transportation, commercial, or industrial activities

9.3.1.18   Oregon (population 2,922,000)

Effective July 1992.
Definition of cleaning agent (in which elemental phosphorus is limited to not more than 0.5%):

> Any product, but not limited to soaps and detergents, containing a surfactant as a wetting or dirt emulsifying agent and used primarily for domestic or commercial cleaning purposes, including but not limited to the cleaning of fabrics, dishes, food utensils and household commercial premises. Does not include foods, drugs, cosmetics, insecticides, fungicides and rodenticides or cleaning agents exempted below.

Exempt:

- Automatic dishwasher cleaning agent containing 8.7% or less elemental phosphorus
- Used in dairy, beverage, or food processing equipment
- Used as an industrial sanitizer, brightener, acid cleaner or metal conditioner, including phosphoric acid products or trisodium phosphate
- Used in hospitals, veterinary hospitals or clinics or health care facilities
- Used in agricultural production and the production of electronic components
- Used in a commercial laundry for laundry services provided to a hospital, veterinary hospital or clinic or health care facility
- Used in industry for metal cleaning or conditioning
- Manufactured, stored or distributed for sale outside Oregon
- Used in any laboratory, including a biological laboratory, research facility, chemical, electronic or engineering laboratory
- Used for cleaning hard surfaces, including household cleansers for windows, sinks, counters, stoves, tubs or other food preparation surfaces and plumbing fixtures
- Used as a water-softening chemical, antiscale chemical or corrosion inhibitor intended for use in closed systems, including but not limited to boilers, air conditioners, cooling towers or hot water systems
- Uses for which the Department of Environmental Quality has determined that the prohibition will either cause a significant hardship on the user or be unreasonable because of the lack of an adequate substitute cleaning agent

9.3.1.19   Pennsylvania (population 11,961,000)

Effective March 1990 for Susquehanna and Erie watersheds, March 1991 elsewhere; sunset December 1992.
Definition of cleaning agent (in which phosphates are banned):

> Any cleaning agent.

Exempt from ban:

- Used in dairy, beverage or food processing
- Industrial sanitizer, brightener, acid cleaner or metal conditioner, including phosphoric acid products or trisodium phosphate
- Used in hospitals, veterinary hospitals or clinics, or health care facilities
- Used in agricultural production
- Used in a commercial laundry that provides services for hospitals, health care facilities, or veterinary hospitals or clinics
- Used by industry for metal cleaning or conditioning
- For use or sale outside the Commonwealth
- Used in any laboratory
- Used for cleaning hard surfaces, including household cleaners for windows, sinks, counters, stoves, tubs or other food preparation surfaces and plumbing fixtures
- Used as a water softening or antiscale chemical or corrosion inhibitor
- Detergent used in a dishwashing machine with no more than 8.7% elemental phosphorus.
- Substance excluded by the Environmental Quality Board by regulation on finding that compliance would create hardship or be unreasonable because of the lack of an adequate substitute

9.3.1.20   *South Carolina (population 3,560,000)*

Effective January 1992.
Definition of cleaning agent (in which phosphates are banned):
A "cleaning agent," meaning a laundry detergent, dishwashing compound, household cleaner, metal cleaner, industrial cleaner, phosphate compound or other substance that is intended to be used for cleaning purposes.

Exempt from ban:

- Cleaning agents used in dairy, beverage, or food processing equipment
- Industrial sanitizers, brighteners, acid cleaners, metal conditioners, including phosphoric acid products or trisodium phosphate
- Used in hospitals, veterinary hospitals, clinics, or health care facilities or in agricultural or dairy production or in the manufacture of health care supplies
- Used by a commercial laundry or textile rental service company or any other commercial entity:
  a. to provide laundry service to hospitals, clinics, nursing homes, other health care facilities, or veterinary hospitals or clinics
  b. to clean textile products owned by a commercial laundry or textile rental service company and supplied to industrial or commercial users of the products on a rental basis
  c. to clean military, professional, industrial, or commercial work uniforms
- Used by industry for metal cleaning or conditioning
- Manufactured, stored, or distributed for use or sale outside the state
- Used in any laboratory
- Used for cleaning hard surfaces, including household cleaners for windows, sinks, counters, ovens, tubs, or other food preparation surfaces and plumbing fixtures

- Used as a water-softening chemical, antiscale chemical, or corrosion inhibitor intended for use in closed systems such as boilers, air conditioners, cooling towers, or hot water heating systems
- By regulations adopted by the Department of Health and Environmental Control based on a finding that compliance would create a significant hardship on the user or be unreasonable because of the lack of an adequate substitute cleaning agent

9.3.1.21   Spokane, Washington (population 177,000)

Effective April 1990, City of Spokane; effective June 1990, Spokane County.
Definition of cleaning agent (in which phosphates are banned):
> Laundry cleaning product. Means any product, including but not limited to, soap, detergent, laundry bleach and laundry additive, used for the purpose of cleaning laundry.

Requires labeling in %P of laundry detergents and soaps whether powdered or liquid, powdered laundry bleaches, and powdered laundry presoak products.

9.3.1.22   Vermont (population 567,000)

Effective April 1978.
Definition of cleaning agent (in which phosphates are banned):
> Any product, including but not limited to soaps and detergents used for domestic or commercial cleaning purposes, including but not limited to the cleansing of fabric, dishes, food utensils and household and commercial premises.

Household cleansing product shall not mean:

- Foods, drugs and cosmetics, including personal care items such as toothpaste, shampoo and hand soap
- Products labeled, advertised, marketed and distributed for use primarily as economic poisons

Completely exempt from ban:

- Household cleaning products used in agricultural production
- Dairy equipment cleansing products
- Exclusion granted on application to the agency of environmental conservation for cleansing products used primarily in industrial manufacturing, production and assembling processes, only if there are no reasonably available alternatives to the user

Elemental phosphorus limitation (8.7%):

- Dishwashers
- Medical and surgical equipment
- Food and beverage processing equipment

9.3.1.23   Virginia (population 6,286,000)

Effective January 1988.

Definition of cleaning agent (in which phosphates are banned):

> Laundry detergent, dishwashing compound, household cleaner, metal cleaner, industrial cleaner, phosphate compound or other substance that is intended to be used for cleaning purposes.

Exempt, but subject to 8.7% [elemental phosphorus] limitation:

- Dishwashers
- Substances for which the Board of Agriculture & Consumer Services finds that compliance would create a hardship or be unreasonable because of lack of an adequate substitute

Completely exempt:

- Dairy, beverage, food processing
- Industrial sanitizers, brighteners, acid cleaners or metal conditioners, including phosphoric acid or trisodium phosphate products
- Hospital, veterinary hospital or clinics, health care facilities, agricultural or dairy production or the manufacture of health care supplies
- Commercial laundry that provides laundry services for a hospital, health care facility or veterinary hospital
- Industry metal cleaners or conditioners
- Laboratory, biological laboratory, research facility, chemical laboratory, and engineering laboratory products
- Commercial laundry that provides laundry services for a hospital, health care facility or veterinary hospital
- Industry metal cleaners or conditioners
- Laboratory, biological laboratory, research facility, chemical laboratory, and engineering laboratory products
- Hard surface cleaners, including household cleaners for windows, sinks, counters, ovens, tubs, or other food preparation surfaces and plumbing fixtures
- Water softening, antiscale or corrosion inhibiting applications
- Products manufactured, stored or distributed for use or sale outside of the Commonwealth

9.3.1.24   Wharton, Texas (population 9000)

Effective October 1992.
Definition of cleaning agent (in which elemental phosphorus in the city of Wharton is limited to 0.5%):

> A cleaning compound in liquid, bar, spray, tablet, flake, powder, or other form used for cleaning purposes.

The term "cleaning compound" shall not mean:

- A metal cleaner, degreasing compound, commercial cleaner, industrial cleaner, or other substance that is intended to be used for industrial cleaning purpose
- A phosphorus [sic] acid product, including a sanitizer, brightener, acid cleaner, or metal conditioner
- A detergent manufactured, stored, or distributed for use or sale outside the city limits
- Dishwashing formulations containing up to 8.7% elemental phosphorus by weight

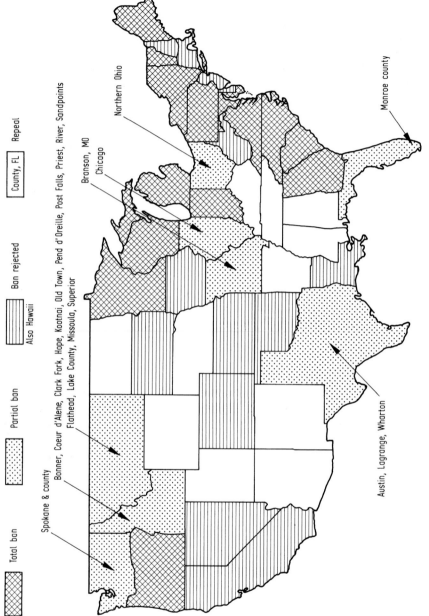

Figure 9.2   Areas of detergent phosphate ban activity in the United States; no such activity in Alaska.

9.3.1.25   *Wisconsin (population 4,955,000)*

Effective January 1984.

Definition of cleaning agent (in which elemental phosphorus is limited to 0.5%):

> Any laundry detergent, laundry additive, dishwashing compound, cleanser, household cleaner, metal cleaner, degreasing compound, commercial cleaner, industrial cleaner, phosphate compound or other substance intended to be used for cleaning purposes.

Completely exempt from ban:

- Industrial processes
- Dairy equipment

Elemental phosphorus limitation (8.7%):

- Automatic dishwashing machines
- Medical and surgical equipment

Elemental phosphorus limitation (20%):

- Chemical water conditioners

Figure 9.2 summarizes, in graphical form, the detergent phosphate ban situation in the United States as of January 1993.

9.3.2   State and Local Phosphate Regulations: Not Enacted

As can be seen from the "rejected" column in the historical bar graph (Fig. 9.1), many bans rejected in a particular area are resubmitted regularly. Accordingly, it is very likely that a state phosphate legislation will be reintroduced in areas where it has been rejected. As of early 1993, detergent phosphate ban activity was possible in Arkansas, Massachusetts, New Hampshire, Texas, and Washington State.

   In some areas, attempts were made to reduce the exemptions of industrial and institutional products from bans. Proposals to lower the 8.7% phosphorus limit for automatic dishwashing products were not adopted because sufficient levels of phosphates are indispensable in these products, as opposed to laundry detergents.

9.4   Results of Phosphate Regulation

9.4.1   Effects on Water Quality

Detergent phosphate bans were enacted on the assumption that removing phosphates from detergents would improve water quality. Water quality of a receiving water is measured by the phosphate level, the chlorophyll (algae) content, the Secchi depth (a visual measure of suspended solids), and the level of dissolved oxygen.

   There exists a popular belief that "every little bit helps," no matter how small, and that any amount of phosphate removed, even if negligible, will still protect and enhance water quality. However, detergent phosphates make a very small contribution ($< 5\%$) to the environment. In addition, as quoted in an American Chemical Society monograph [6], independent of any detergent source, average domestic sewage contains sufficient uncontrollable phosphorus to support the growth of 1 g/L of blue-green algae, 50 times the algal concentration found in Clear Lake, California; 100 times that found in Lake Erie; and 1000 times that found in oceans.

Several U.S. studies are relevant. After 18 months of Indiana detergent ban in the White and Wabash rivers, phosphate levels failed to improve [7]. A study performed in 1989 at Purdue University [8] showed no improvement in water quality (chlorophyll, Secchi depth, total phosphorus) of 14 out of 15 lakes after 16 years of detergent bans in Indiana. Similar results were obtained in a seven-lake Wisconsin study by the Wisconsin Department of Natural Resouces after 3 years of bans and after 11 years [9]. A 1986 modeling study [10] indicated that none of Wisconsin's 15,000 lakes would show a visible change from the ban. A Minnesota study on six pairs of lakes found no effect of that ban on Minnesota water quality. The Virginia Water Control Board [11] confirmed that there was no evidence of water quality improvement in Indiana, in Vermont, or in Wisconsin attributable to detergent bans in these states. In September 1988, the state of Maryland [12] confirmed that even though the law required that ban results be documented, the Department of the Environment would be unable to comply. In Pennsylvania, three studies calculated the phosphorus reduction in the Susquehanna River (the main freshwater tributary into the Chesapeake Bay) that a detergent phosphate ban would cause. One report [13] set this reduction at 1.5–2.8%; another, from The Pennsylvania State University [14], found 3.2–4.0%; and the EPA calculated 2.1%.

Several reports describe the Great Lakes in general. Nonpoint sources accounted for 85% of the total phosphorus contribution to Illinois lakes [15]. Of the total phosphorus entering the Great Lakes, only 14% resulted from municipal discharges [16]. While the Great Lakes water quality has indeed improved since the 1970s, this is not due to detergent bans. The U.S. Army Corps of Engineers [17] found that phosphorus loadings decreased from about 20,000 to 16,500 metric tons per year but attributed this result to the construction of large municipal treatment plants. Another independent report confirmed these findings [18]. A U.S. National Oceanic and Atmospheric Administration report confirmed that detergent phosphate bans could not improve the quality of the Great Lakes [19]. An EPA study of 493 lakes across the United States concluded that a ban would have "very little effect on lake water quality" [20]. Recent studies in Britain, Italy, and Switzerland have confirmed these results.

A final factor is that the presence of phosphates, even in large amounts, is often of no consequence in some environmental systems. In marine waters, for example, nitrogen is generally the nutrient present in the lowest relative amount [21], so that phosphorus has no effect. Bans assume that phosphorus is the limiting element for algal growth, but it has been shown that in 1260 Lake Erie algal cultures, nitrogen was growth-enhancing twice as often as phosphate [6]. For eutrophication to occur, each link in a chain of events must be favorable: vital elements and some vitamins must all be present [22]. In addition, temperatures, sunlight, predators (e.g., grazing zooplankton), and parasitism must all be favorable. If but a single one of these many factors is unfavorable, no amount of phosphate will cause algae to grow.

9.4.2   Effects on Sewage Treatment Plants

On a local scale, detergent phosphate bans have been effective in reducing the phosphate outflow of sewage treatment plants. Typically, this enables some plants to relax standards, or it reduces the cost of the chemicals used in phosphate removal operations. The annual savings for all U.S. sewage treatment plants have been estimated at $2-3 million [23].

9.4.3   Other Effects of Phosphate Bans

Some recent reports blame the *lack* of phosphates for unfavorable environmental effects. Toxic red tides are bursts of growth by different species of algae that can transmit poisons to people through "filter-feeding" shellfish, which strain nutrients from seawater and concentrate toxins in their internal organs. Increases in toxic red tides have been reported. Dr. Donald M. Anderson, of the Woods Hole Oceanographic Institute in Massachusetts, suggested that phosphate-free detergents may contriubte to red tides by changing chemical ratios (reflecting the lack of phosphate) in coastal waters [24]. A scientific publication indicates that under culture conditions *deficient in phosphate,* dinoflagellates (*Alexandrium spp.*) become more toxic by continuing to produce saxitoxin, despite being slowed in growth [25].

While living algae are aesthetically unappealing, they fix large amounts of atmospheric carbon dioxide, some of it anthropogenic, which may contribute to the *greenhouse effect* [26]. The mass balance equation [27] for the growth of algae under phosphate-limiting nutrient conditions,

$$550\ CO_2 + 580\ H_2O + 30\ HNO_3 + H_3PO_4 \rightarrow (CH_2O)_{550}(NH_3)_{30}(H_3PO_4) + 610\ O_2 \qquad (9.1)$$

shows that the phosphate equivalent to one pound of phosphorus, discharged into a surface water and entirely converted to living algal tissue, will fixate 350 kg of CO_2 or 179,000 L at standard temperature and pressures. Environmental concern about a possible greenhouse effect was unknown in the 1970s when detergent phosphate regulation originated. Global warming has become a concern today. Algae appear to be of importance in alleviating it.

In the Netherlands, the Ministry of Fisheries [28] claimed that the level of various fish and shellfish in the Rhine had fallen to 50% of its former value as a result of phosphate reductions in that river. This effect is reminiscent of the inability of any fish to live in Lake Mead without added phosphate, quoted earlier [7].

9.5   Surfactant Biodegradation

The foaming of early surfactants in the environment has a highly interesting history. In the United States and Canada, this problem was settled totally without regulation.

Early detergents were based on soap and presented many problems of both performance and availability. In 1930, the German firm of H.T. Bohme developed and marketed synthetic alcohol sulfates (Gardinol), while I.G. Farbenindustrie, also in Germany, developed the Igepons, originally taurine amide materials. Both product lines were marketed in the United States in the early 1930s, being first used in shampoos and "soapless" detergents which, while popular, did not exhibit very good soil removal. Wartime research into saltwater detergents produced the classic synthetic detergents containing sodium tripolyphosphate as builders and branched tetrapropylene arylsulfonates (ABS) as anionic surfactants. Introduced in 1946 by Procter & Gamble as Tide, synthetic detergents of this type quickly surpassed all soap products, and by the early 1950s had acquired the lion's share of the market.

Shortly after the introduction of built detergents, foaming was detected at sewage treatment plants—and later at home faucets. Industry and government quickly identified the cause of this foam as the poor biodegradability of the ABS surfactants [29]. Beginning in 1961, regulations passed in West Germany mandated surfactant decomposability, and much similar legislation was introduced in the United States, but none of these measures were enacted. The problem was quickly recognized by the industry, and a plan was developed to solve it by substituting ABS with linear alkylbenzenesulfonate (LAS) [30]. LAS is a material that biodegrades satisfactorily and does not cause any environmental foam or other residual symptoms.

- ABS simplified: $CH_3[CH(CH_3)CH_2]_3CH(CH_3)C_6H_4SO_2ONa$
- LAS simplified: $CH_3[CH_2]_{11}C_6H_4SO_2ONa$

After prolonged hearings and testimony, LAS made its commerical appearance in 1966. The problem was solved and has not recurred. This is a good example of the achievement of a higher quality environment by cooperation among regulators, industry, and scientists. There was no sensationalist publicity generated by adversarial advocacy.

Some states did indeed enact legislation, but it was of no regulatory or problem-solving relevance. In April 1963, for example, the Dade County, Florida, Board of Commissioners enacted a local ordinance later amended into the Dade County phosphate ban (repealed in 1985), which prohibited the sale, use, or possession of detergent soaps [sic] or biologically nondegradable detergents [sic]. In 1964, a similar bill was enacted in Wisconsin, banning the sale of nonbiodegradable household detergents effective January 1966. Similar bills were introduced but not passed in Nebraska, California, and Connecticut. The Indiana phosphate ban also prohibits the sale or use of detergents containing ABS. A 1971 ordinance in Bayville, New York, prohibited ABS, alkyl sulfates (AS), methylene blue active substances (MBAS), and nonionic surfactants in detergents. A 1971 ordinance in New Shoreham, Rhode Island, prohibited ABS, MBAS, and nonionic surfactants in detergents.

The reader is referred to Section 9.3 on U.S. state and local detergent phosphate regulation for details of some of the legislation above.

9.6   pH Regulation

A Milwaukee city ordinance of December 1978 prohibits the sale of any household laundry detergent which, based on a 1% solution in distilled water, has pH higher than 10.9.

9.7   NTA Regulation

A 1971 regulation in Kennebunkport, Maine, limiting phosphate in detergents to 14% banned the presence of nitrate or nitrilotriacetic acid (NTA).

9.8   Waste and Solvent Regulations

9.8.1   California

California Proposition 65 (California Safe Drinking Water and Toxic Enforcement Act of 1986) provides that no person in the course of doing business shall knowingly discharge or release any detectable amount of a chemical known to the state to cause cancer or reproductive toxicity. "Discharge" is defined as "into water or unto and into land where such chemical passes or probably will pass into any source of drinking water." In addition, Proposition 65 forbids all persons in the course of doing business to knowingly and intentionally expose any individual to a such a chemical without first giving clear and reasonable warning. The California EPA, specifically the Office of Environmental Health Hazard Assessment (OEHHA), is the leading agency that will determine which chemicals are carcinogenic or reproductive toxicants. OEHHA will also determine risk levels that require hazard notification under Proposition 65.

The important philosophical/mathematical dispute about the meaning of "any detectable amount" and "zero" is an active topic, especially in connection with the Delaney Clause of the Food, Drug, and Cosmetic Act. Enacted 35 years ago, the Delaney Clause outlaws any detected amount of a human or animal carcinogen in processed food. The minimum amount detectable clearly changes with improvements in analytical techniques. Carcinogens, metals, and all other materials eventually will be found in samples in which they are now thought to be "zero" because they are below the detection limit. In a 1988 EPA interpretation reendorsed in 1993, EPA held that the Delaney Clause exempts pesticide residues that are small enough to present only a negligible health risk in processed foods. "Negligible risk" was defined as one additional cancer case per million people resulting from a 70-year exposure. Environmental groups contended that U.S. laws had failed to protect public health and that only "zero"-risk definitions were acceptable. A 1992 appeals court ruling overturned the EPA interpretation, but the decision is not being applied pending news of whether the U.S. Supreme Court will review the ruling.

A related factor is the magnitude of desirable safety factors (maximum no-effect concentration divided by existing concentration), such as 1:10 and 1:100. Recently, safety factors that many materials in nature or in industrial products could not meet have been proposed, some as high as 1:1000.

California's Proposition 65 requires the governor to revise and publish the list of chemicals known to the state to cause cancer and reproductive toxicity. Any of these chemicals in a consumer product is the subject of point-of-sale warnings that are inflexible as to content. An update of October 1992 includes a number of well-known drugs such as aspirin, all barbiturates, actinomycin, benzodiazepines (Valium-type drugs), diazepam, fluorazepam, oxazepam, ethyl alcohol in beverages, neomycin, tetracycline, and oxytetracycline [31]. In the cleaning product area, ethylene oxide, benzene, NTA, and arsenic are on the list.

In addition, California Proposition 105, which directs a very stringent assessment of the toxicological and environmental merits of commercial products, went into effect in January 1990.

In 1992, Ohio considered and rejected Proposition 5, a substantially more restrictive analog of California Proposition 65. The measure required each manufacturer to conduct its own risk assessment on all products and then hire an outside risk assessor to assess the assessment.

The federal preemption doctrine, which is based on the supremacy clause (Article VI, clause 2, of the Constitution), holds that state statutes, regulations, and even state common law are precluded by federal law under certain circumstances. Federal preemption of state and local law regarding the labeling and warnings on chemical products is one case of such an override. The preemptive effect of the Federal Insecticide, Fungicide, and Rodenticide Act (FIFRA) and the Federal Hazardous Substances Act (FHSA) has thus been litigated [32] with respect to the point-of-sale warnings mandated by Proposition 65. However, the appeals court found that if Proposition 65 were preempted under FIFRA or FHSA, the court would have to determine that all possible consumer product warnings that would satisfy the state statute conflicted with FIFRA and FHSA provisions. The U.S. Supreme Court announced in October 1992 that it would not consider a petition seeking review of this ruling.

9.8.2   Florida Solvents Tax

The State of Florida levies a solvents tax on consumer and institutional products to fund state Superfund cleanup operations and restoration or replacement of private potable water wells. This tax is $0.02–0.03 per gallon, and 1992 efforts, likely to be retried in 1993, sought to raise the tax to $0.24/gal. In 1992, the Florida Department of Revenue discontinued the exemption from this tax that detergents had enjoyed.

9.8.3   Volatile Organic Compound Regulation

Volatile organic compounds (VOCs) are believed to contribute to amphoteric ozone formation, and federal ozone standards are not being met in approximately 50% of the United States. VOCs are sweepingly defined. Many laundry and hand dishwashing detergent and other cleaning agent ingredients may qualify, by this elastic and imprecise definition, even though they do not find their way into the atmosphere. Low molecular weight alcohols are definitely VOCs.

At the federal level, EPA has recommended a national inventory of VOC emissions, including assessments of the fate of VOCs in discarded consumer products.

The California Air Resources Board recently altered regulation that would have imposed VOC content limits on laundry detergents and hand dishwashing products. Regulations limiting VOC content of certain product categories, including hard-surface cleaners, now exist in California and New York. Texas also has regulations in this area. New Jersey has mandated VOC record keeping for manufacturers of consumer products containing more than 5% VOCs. In addition, the Ozone Transport Commission will examine consumer products for VOCs in 1993.

9.8.4   Household Hazardous Waste Regulation

A number of states have passed laws aimed at managing household hazardous waste. They mandate a search for, and information programs on, nontoxic alternatives, proper use, storage, disposal procedures, and recycling options. The general criterion for a hazardous household waste is usually the presence of a constituent that makes disposal in municipal waste landfills or incinerators undesirable for reasons of causticity, corrosivity, flammability, irritancy, reactivity, toxicity, and so on. The actual definition of what household products generate hazardous waste is, however, quite variable. Thus, a recent Illinois statute [33] includes waste oil, herbicides, pesticides, solvents, liquid paint, paint remover, and paint thinners but also characterizes disinfectants, and possibly other cleaners, as hazardous. Disinfectant products are considered to be pesticides under FIFRA.

The broad clauses on nonhazardous alternatives themselves constitute a full regulatory system. Thus, the California EPA has distributed a brochure on "hazardous household products" that includes household cleaning products and recommends "safer practices" (borax, baking soda, vinegar, salt, club soda, lemons, and toothpaste). It also includes all paints, stains, varnishes, photographic chemicals, pool chemicals, glues, antifreeze compounds, batteries, brake fluid, drugs, all medicines, shoe polish, and "chemical" fertilizers.

9.9 Trends in the United States

9.9.1 Animal Testing Regulations

Animal testing is of only marginal relevance to detergent regulation. There has been much regulatory and research activity on alternatives to animal testing and the Draize eye irritation test. Licensing regulations of such testing exist in New York and have been proposed but not adopted in some six other states.

9.9.2 Fragrance Regulation

Since popular lore considers a fragrance more a "chemical" than a "product," some local regulations on fragrance limitations, including indeed a total ban on all fragrances, have been proposed. As of this writing, no such regulation has been adopted.

9.9.3 Media Image: Eco-Marketing

Strong public opinion, the precursor of regulation, exists today regarding chemicals and toxicology. The media and the public have simplified toxicology into quick judgments as to what is "good" and "bad" for you. Even though the most potent carcinogen (aflatoxin) and the most potent poison (botulism toxin) are natural products, public opinion holds that natural products are not chemicals and are generally "good" for you. An example in the cleaning products area is vinegar. White vinegar supermarket sales have increased more than 20% since 1986. Consumer advocates believe that vinegar is not a chemical because it is made by fermenting such natural materials as wheat, grapes, apples, or rice. Not only is vinegar reported to clean stubborn yellow stains on a divan that all the "advertised cleaners of corporate America" did not remove, but it is also purported to have hundreds of other cleaning uses.

This mind-set leads to "eco-markeing": the term "natural" is ill-defined and is thus easily applied to the origin of products. In addition, commercial and advertising claims such as "chlorine-free," "ozone-friendly," and "no animal testing" have been used to launch cleaning products, especially in several European countries and Canada. The specific thrust is similar for eco-marketing and for regulation, since consumerist and other groups would like to see their recommendations enacted as laws. The disproportionately positive image of natural products has exerted a *de facto* regulatory effect on both the formulation and the use of cleaning products.

Eco-marketing extends to many other cleaning product applications. In some cases, it has led to reduced use of cleaners in hotels and to hotel towel changes made only upon customer request. Other instances include watering of flowers with well or ditch water, composting of all hotel waste, use of biodegradable popcorn in pillowed packaging, promotion of solar energy, and recycling.

Likewise, image considerations find their way into regulatory action. A typical example is a 1992 CERCLA Superfund amendment according to which "household products" are assumed to be "hazardous." This amendment would have required municipalities to implement a public education program "that identifies both hazardous household products and safer substitutes."

9.9.4   Certification Programs

A number of environmental and nonprofit organizations (Green Cross, Green Seal [34], et al.) have been working on product standards that would enable the award of approval certificates and/or seals for easy public identification of claimed environmental benefits. Some of these are based on arbitrary standards; others mandate full and extensive life cycle analyses. Several trade associations are also working on product standards with a similar thrust.

9.9.5   Definition Problems

The absence of clear definitions for many of the concepts used in eco-marketing and certification programs is a perplexing difficulty. An important example is the widely used term "biodegradability." Section 9.2.2 describes FTC efforts to address guidelines and examples, although not definitions. Besides the FTC, many other bodies are independently attempting to define "biodegradable" and the determine the relative importance of aerobic and anaerobic events. These include the American Society for Testing and Materials (ASTM) [35], the Chemical Specialties Manufacturers Association (CSMA), the State of California, Canada, the U.S. EPA, a variety of European groups [36], and several groups in the Soap & Detergent Association (SDA). In spite of this multiplicity of efforts, no clear consensus, hence no single definition of the term, has been reached.

Likewise, many other terms (toxic, hazardous, safe, pollutant, friendly, harsh, etc.), all vitally important in the description and the claims regarding detergents and cleaning agents, have no generally agreed-on definition. They are thus prone to misuse and misunderstanding. Often, such a word will be defined in one way in common usage and in another way in law. Finally, it will then be used, usually in advocacy of regulation, in a third frame of reference. Grammar, law, and politics then are at odds. People think they know what is meant, yet there is no consensus.

9.10   Canada

9.10.1   Phosphate Regulation

Previously under the Canada Water Act (Loi sur les Ressources en Eau du Canada), Section 50 of the Canadian Environmental Protection Act or CEPA (Loi Canadienne sur la Protection de l'Environnement), effective October 1989, regulates the concentration of phosphorus in any laundry detergent (détergent à lessive) not to exceed 5% by weight expressed as phosphorus pentoxide or 2.2% by weight expressed as elemental phosphorus [37]. Analytical methods for the determination of these percentages are specified by law [38]. Compliance will be verified by inspectors, and the CEPA provides for the assignment of penalties for contravention or failure to comply.

9.10.2   Surfactant Regulation

Canada and its provinces do not actively regulate surfactants.

9.10.3   Right-to-Know Law

In September 1992, the Canadian government announced an annual toxic chemical emission reporting system similar to that of the United States. Modeled on the U.S. Toxic Release Inventory (TRI), the Canadian system will be called the National Pollutant Release Inventory (NPRI). First reports are due in April 1994 for the 1993 year. The Canadian definition for toxicity is unknown. (The definition in the corresponding U.S. law includes substances that cause an adverse effect to the environment because of their acute or chronic toxicity to living systems.)

References

1. Axler, R.P., Paulson, L.J., et al., Lake Reservoir Manage., *4*, 125 (1988); also Lake Mead Fertilization Project Report, University of Nevada, May 1987.
2. Warren, C.E., *Biology and Water Pollution Control,* Saunders, Philadelphia, 1971.
3. Fed. Regis. 57(157) 36363–36369 (Aug. 13, 1992).
4. U.S. Senate, Congressional Record, S5786–S5790, May 14, 1991.
5. *Statistical Abstract of the United States, 1992,* Table 25. U.S. Dept. of Commerce, Bureau of the Census, Washington, DC.
6. Toy, A.D.F., Walsh, E.N.,*Phosphorus Chemicals in Everyday Living*, American Chemical Society, Washington, DC, 1987.
7. Etzel, J.E., Bell, J.M., Water Sewage Works, *9*, 91 (1975).
8. Spacie, A., Loeb, S.L., Verh. Int. Verein. Limno., 24, 464 (1990).
9. State of Wisconsin, Bureau of Water Resources Management, Dept. of Natural Resources: Report to the Legislature on the Effects of the Phosphate Ban, Madison, WI 53707, January 1988.
10. Clesceri, N., Sedlak, R., Water Res. Bull., *22*, 6 (1986).
11. Virginia State Water Control Board Phosphate Detergent Ban Task Force, Report to the Chesapeake Bay Commission, State Water Control Board, and Virginia Cooperative Extension Service, Annapolis, MD, November 1984.

12. Hearn, J.L., Department of the Environment, State of Maryland Report on Ban Results by the Director of Water Management Administration, Maryland, September 26, 1988, Baltimore, MD.
13. Pennsylvania State University, Env. Resources Res. Institute, Dymond, R.L., Long, D.A., Assessment of the Impact of Phosphate Detergent Ban, Report ER8803 to Chesapeake Bay Commission, March 1988, University Park, PA.
14. Dymond, R.L., Long, D.A., Pennsylvania State University, March 1988.
15. Garman, G.D., Good, G.B., Hinsman, L.M., Summary of Information Regarding Lake Water Quality, Illinois Environmental Protection Agency, 1986, Illinois EPA, Div. Water Poll. Ctrl., Springfield, IL.
16. Great Lakes Water Quality Board Survey Report to the International Joint Commission, 1981, Windsor, ON N9A 6T3, or Detroit, MI 48232.
17. Lake Erie Waste Water Management Study, 1983, U.S. Army Corps of Engineers, Buffalo District, Buffalo, NY 14207.
18. Lee, G.F., et al., Environ. Sci. Technol., *12*, 900 (1978).
19. Lee, G.F., Jones, R.A., National Oceanic and Atmospheric Administration Report, 1979, Colo. State University, Ft. Collins, CO.
20. Lorenzen, M.W., U.S. Environmental Protection Agency document no. 560/11-79-011, 1979, Washington, DC 20460.
21. Ryther, N., Dunstan, Science, *171*, 1008 (1971).
22. Hutchenson, G.E., *A Treatise on Limnology,* Vol. 2, Wiley, New York, 1967.
23. Clayton Associates, Potential Environmental and Economic Benefits of Discontinuing the Use of Phosphates in Laundry Detergents, Draft Report to the U.S. Environmental Protection Agency, 1992.
24. Wall St. Journal, Nov. 22, 1991.
25. Anderson, D.M., Mar. Biol., *104*, 511–524 (19__).
26. Post, W. M., et al., Am. Sci., *78*, 310 (1990).
27. Atkinson, M.J., Smith, S.V., Limnol. Oceanogr., *28*, 568–574 (1983).
28. Boddek, D., Dutch Royal Institute of Research, Feb. 29, 1992.
29. McGucken, W., *Biodegradable: Detergents and the Environment,* Texas A&M Press, College Station, 1991.
30. Swisher, R.D., *Surfactant Biodegradation,* Marcel Dekker, New York, 1987.
31. California Regulation Notice Register, 1992, pp. 1234–1244.
32. Chem. Times. Trends, *15*(4) 58 (1992).
33. Illinois Public Law 87-1159, H3029, 1992.
34. Proposed Environmental Standard for the Certification of General Household Cleaners, Green Seal, Aug. 1992, Washington, DC 20037.
35. American Society for Testing and Materials, Committee Reports on the Biodegradability of Synthetic Materials; Biological Effects and Environmental Fate; Environmental Fate and Activated Sludge; Terminology: ASTM, Philadelphia.
36. European Community Directive 67/548, EEC, Brussels, Belgium.
37. Canada Gazette, Part II, Vol. 123, No. 23, Aug. 11, 1989.
38. Department of Environment, EPS 4-WP-74-2; EPS 1-WP-76-1, Ottawa, Canada.

Bibliography

Atkinson, M.J., Smith, S.V., C:N:P ratios of benthic marine plants, Limnol. Oceanogr. *28*, 568–574 (1983).
Canada Gazette Part II, Vol. 123, No. 23, Aug. 11, 1989, SOR/DORS/89-501.
Code of Federal Regulations (CFR) Update Service, yearly, Government Institutes, Rockville, MD 20850.
Directory of Environmental Information Sources, 4th ed., ISBN 0-86587-326-7.
Environmental Law Handbook, 11th ed., ISBN 0-86587-250-3.
Environmental Phosphorus Handbook, E.J. Griffith, Ed., Wiley, New York, 1973, ISBN 0-471-32779-4.
Environmental Regulatory Glossary, ISBN 0-86587-798-X.

Environmental Statutes, 1992 ed., Government Institutes, Rockville, MD, ISBN 0-86587-282-1.

Fundamentals of Environmental Science and Technology, P.C. Knowles, Ed., ISBN 0-86587-302-X.

Green Seal, Proposed Environmental Standard for the Certification of General Purpose Household Cleaners, Aug. 1992.

Jenkins, D., Nitrogen and phosphorus forms in natural waters, in *Trace Inorganics in Water,* Robert A. Baker, Ed., American Chemical Society, Washington, DC, 1968.

Lee et al., Environ. Sci. Technol., *12*, 900 (1978).

Management strategies for phosphorus in the environment, in *Proceedings of International Conference,* Lisbon, July 1–4, 1985, Selper, London, ISBN 0-948411-00-7.

Rogers, W.H., Jr., *Handbook of Environmental Law,* West Publishing Co., St. Paul, MN, 1977, ISBN 0-314-33231-6.

Spacie, A., Loeb, S.L., Effects of detergent phosphate ban in Indiana, Verh. Int. Verein. Limnol., *24*, 464–469 (1990).

State Environmental Law Handbooks, Government Institutes, Rockville, MD.

Statistical Abstract of the U.S., 1992, Table 25.

Toy, A.D.F., Walsh, E.N., *Phosphorus Chemistry in Everyday Living,* American Chemical Society, Washington, DC, 1987, ISBN 0-8412-1002-0.

U.K. Department of the Environment, Pollutants in Cleaning Agents, Mar. 1991; see also AIS/FIFE newsletter Info-Bulletin, Sept. 1, 1991.

Warren, C.E., *Biology and Water Pollution Control,* Saunders, Philadephia, 1971.

CHAPTER 10

Hints for the Formulator

K. Robert Lange

The formulator is often faced with problems that impinge on product quality. These include formulation stability, appearance, odor, and corrosivity. Problems, which exist with both liquid and solid products, often are quite subtle, and the solutions are by no means universal. The formulator must call on general knowledge of chemistry, ingenuity, and, above all, experience.

This chapter discusses the problems just named for solid products, solutions, slurries, and emulsions. In addition to the problems associated with product stability and appearance, production problems are discussed, especially for batch processing.

10.1   Introduction

The formulator would do well to keep the history of detergency in mind in the course of product development. Here water quality assumes an important place in the scheme of things. The substitution of synthetic wetting agents for soap took place largely because of water hardness, which causes insoluble Ca and Mg soaps to precipitate, forming a floating scum that is deposited on laundry and equipment. Synthetic linear alkylbenzenesulfonate (LAS) detergents do not form these insoluble salts, and the absence of such materials results in cleaner conditions. More recently, the nonionic surfactants made possible lower wash temperatures. Amphoterics allow the formulation of shampoos that do not sting the eyes. Enzymes are today providing greater selectivity regarding soils. Thus, water quality, energy consumption, and human needs are among the factors driving product development. Now the field has been expanded, with societal and environmental requirements assuming center stage. Meeting all these needs is not easy, inasmuch as product complexity increases with the new performance criteria being expressed.

This chapter discusses some problems frequently encountered when putting products together. It is not sufficient to simply pour all the ingredients into a beaker and declare the mixture a product. Subsequent testing may reveal instabilities of various kinds, such as splitting or drop-out. There may be other deficiencies, such as too much foam, a bad odor, or long-term instability not easily noted in the laboratory. Corrosion of containers or machinery also presents a serious defect in a product. Unless, moreover, the product is easy to manufacture, a problem costing time and money may result.

Prescribing the proper chemicals is only half the battle. Some applications demand pastes or solids; others need easily pumped solutions. All must have their active components at the proper concentration for dosing into the system. It is not possible to present here all the

---

K. Robert Lange, 805 Lombard Street, Philadelphia, Pennsylvania 19147, U.S.A.

problems that can occur, or all the possible solutions. This chapter is intended to stimulate thought and, thereby, action by the formulating chemist. The task consists largely of foreseeing problems and defining them as they arise.

10.2   Solid Formulations

Mixing solids sounds easy. They will not react with each other, but will simply sit there until used . . . or so one might think. Problems with solid formulations are, unhappily, plentiful, as discussed in Sections 10.2.1–10.2.5.

10.2.1   Settling or Segregation

Even low degrees of vibration during shipping and handling may cause segregation of well-mixed ingredients. The two main causes of separation are density and particle size. As to density, it is seldom the true density (handbook value) that is to be considered. Rather, the bulk density comes into play. This property has become more and more important as raw materials have been made available in fluffed, dendritic, and porous forms. Particle size assumes significance when powders of widely different sizes are used and the smaller particles sift downward.

Segregation is not always easily detected if, let us say, the ingredients are all white and similar in particle size. First, test for segregation. One way is to color each ingredient differently, using vegetable dyes, for example. then place the jar on a vibrating platform and examine it after an hour. The vibrating table should, in a short time, simulate shipping and other handling effects on the product. If segregation has occurred, most probably the densest and/or the finest ingredients will have dropped to the bottom, leaving the lightest or largest on top.

Look at the data available on bulk densities and particle size of the ingredients. Examine alternatives:

- Are these particles anhydrous and are hydrated forms available with lower densities?
- Could a more porous form be obtained? Such products are available as spray dried versions most frequently.
- Is a dendritic form available? A product that has been crystallized in a form having thin branches (dendrites) emanating from a central nucleus, rather than as discrete dense crystals, will be lower in bulk density.
- What about substituting another chemical that has the same functionality but is better suited for the formulation? Many alternatives exist and should be explored.

A related problem is product compaction. Customer complaints abound, particularly with consumer products, of packages only partly filled. Such units were filled completely during packaging: compaction is responsible for the empty area in the containers. The use of low bulk density raw materials, where available, can alleviate this problem. Some suppliers print explanations on the package, but this is less desirable than overcoming the basic problem.

Our discussion has, so far, considered simple mixtures. Clearly, for a manufacturing facility that has the equipment, agglomeration methods will solve many of the problems mentioned. In these methods, dry powders are mixed and a liquid ingredient added, often by spraying, to agglomerate the mixture. The liquid used may be sodium silicate solution, a polyacrylate solution, or some other functional material in liquid form. Dryness may be achieved through the hydration of ingredients or by evaporation. Sometimes it is possible to react with water in a more sophisticated manner. For example, several years ago a cyclic phosphate was introduced that reacted with water, opening up the structure. The result was a dry, agglomerated mass. While that particular product did not find acceptance, it did point the way to a useful technique. For more on agglomeration methods, see Laundry Products (Chapter 6), which also discusses spray-drying, a widely used method to achieve dry, fine powdered products starting from a slurry, rather than a physical mixture.

10.2.2   Caking

After some time in storage, solid mixtures may compact into masses that need to be broken up mechanically. This is commonly called caking (when it happens to sheets of material, the term blocking is used). Whatever its name, such material behavior is a frustrating problem to the customer and the supplier.

First, the reason for the problem needs to be found. Usually caking is due to the presence of water in some form—not only bulk water added to the product but possibly water absorbed from the air. Or, water of hydration from one ingredient may have been "stolen" by another, which has a higher affinity for water. Other fluids added to the formula may be at fault, but the problem is usually water.

Examine the ingredients, looking for hydrated and anhydrous salts. Assume that a hydrated salt can lose water to an anhydrous salt and look for alternatives. Keeping the system totally anhydrous, or totally hydrated, may be the answer. A completely anhydrous system may deliquesce, absorbing moisture from the air. A hydrate may even do so; calcium chloride is one example. Juggling the various combinations and examining the results after storage should provide the answer. Storage tests may be open to the air or, preferably, conducted under known, constant humidity conditions, in a closed chamber.

Some ingredients have a high affinity for water but do not contribute to caking. Here high surface area, hydrophilic solids such as clay or finely divided (precipitated) silica offer the best prospects. Small amounts can do the job, often 1% or less. Choose solids with 200 m^2/g or higher surface area, and be aware that various forms are available. Clays include bentonite, kaolin, and montmorillonite. Silicas exhibit various properties depending on their manufacture. Fumed and precipitated silicas have a highly particulate nature and make excellent flow control aids for solids. Precipitated silicas generally have a higher affinity for water than fumed silica. Investigate several alternatives.

10.2.3   Color

Some rather subtle color problems can arise. For example, nonionic alkylphenol ethoxylate (APE) surfactants in the formula may turn red or purple because quinoid structures are present in trace amounts. In this case, look for sources of alkalinity, keeping in mind that quinoids develop color generally with elevated pH and that water aids color development. Therefore eliminating sources of moisture and strong alkaline ingredients (e.g., NaOH) can prevent this problem. It may be possible to substitute sodium carbonate for sodium hydroxide. Encapsulating the alkaline ingredients is an expensive but sometimes feasible course of action. Should such an approach fail, seek another source of nonionic surfactant, since manufacturing controls differ among suppliers.

Optical brighteners may turn yellow or pink when in contact with alkali. This effect is due to increased resonance caused by the abstraction of labile hydrogens and the formation of azo nitrogens, as in the case of 4,4'-diaminostilbene derivatives. Here the labile hydrogens attached to the amino nitrogens can be abstracted by base, extending the molecule's resonance to the triazinyl groups, with a shift toward the longer wavelength end of the spectrum. This completely reversible reaction does not interfere with functionality, but such knowledge is beside the point when product appearance is compromised.

Substituting an optical brightener having no labile hydrogens is a very real alternative when a problem such as the diaminostilbene situation referred to above occurs. The labile hydrogen can be replaced with a methyl group, for example. Or an optical brightener of another chemical class might be substituted, such as a coumarin derivative. Coloring the entire product may be one way to correct the situation, and the color itself may have marketing appeal. See Chapter 4 (Section 4.2) for additional information on optional brighteners.

In any product, gradual color development during storage should signal a slow chemical reaction, possibly oxidation. Any such observation needs to be investigated to ensure that functionality and product safety have not been compromised.

10.2.4   Corrosion

Even solids can be corrosive, generally because of the presence of water, as water of hydration, as condensed atmospheric moisture, or absorbed from the air as the result of deliquescence. This problem usually appears first in manufacturing equipment and containers used for shipping the product, such as steel drums.

Changing to plastic or fiber containers is probably the most effective means of eliminating steel corrosion. If there should be an overriding reason to stay with steel, it is best to keep in mind that steel corrosion is at its lowest at about pH 8.5. Below that value FeII and FeIII are formed, while above 8.5 ferrites and ferrates are the result. If an additive is called for, silicates are effective, particularly those termed "water glass," having SiO_2/Na_2O ratios of 2.4–3.2. The silicates may be incorporated as a powdered glass, spray-dried hollow beads, or even as the liquid if there is some adsorptive material present, such as clay, 1–3% should be sufficient.

Since cleaners are seldom at a neutral pH, aluminum, zinc, or their alloys should not be used. Plastic or rubber coatings may be suitable, however.

10.2.5   Dusting

The generation of dust by a solid product can present a frustrating problem. Dusts can be hazardous, causing irritation of the eyes and throat and, if inhaled, of the bronchial passages. Organic dusts (e.g., flour) can cause explosions. The generation of excessive amounts of dust must be avoided.

The most common approach is to add a small amount of a nonvolatile liquid to coat the offending particles. Nonionic surfactants, paraffin oil, glycols, poly(ethylene glycol), and so on have all been used in this connection. An approach that can be productive is the use of pine oil or D-limonene, each of which acts as a cleaning agent and has a distinctive and pleasant odor. Pine oil is commonly used in liquid products so that its odor is associated with cleanliness. Limonene has a citrus odor, which also is "clean."

Mineral oils, though generally good as dust inhibitors, should not be used on some products. For example, electrolytic cleaners, if treated with oil to suppress dust, will suffer in the application if the oil floats in the cleaning bath, coating the metal and interfering with the electroplating action. Uneven or poorly adherent plating may result.

10.3   Liquids

In this section we concentrate on water as the primary solvent used, naturally. Whenever water is employed for product development its quality must be taken into account. The main attributes in this respect are pH and water hardness. Most water has a pH of about 6.5, although in some well and reservoir waters the alkalinity may be higher. Water hardness is due to dissolved ions, specifically Ca and Mg. Hardness is commonly expressed as parts per million and stated as $CaCO_3$. Municipal water can range in hardness from about 50 ppm to 250 ppm or even higher.

Water quality impinges on products, since hardness ions will react with dissolved ingredients to form insoluble salts. Silicates and orthophosphate are examples of this problem, but many other salts are subject to insolubilization. Both hardness and pH are factors in emulsion stability. High ionic strength and improper pH tend to destabilize emulsions.

Knowing water quality, and taking corrective action, is therefore important. Industrially, the alternatives are to specify water that has been either deionized or softened. The former is preferred, since softening, while removing the hardness ions, retains the problems associated with ionic strength. Generally softened water can be used for solutions, but deionized water must be used for emulsions.

It must be pointed out that customer dissatisfaction with products may also be the result of water quality, whether the customer is industrial or in the home. Home problems are well recognized by detergent suppliers. Industrial problems are aggravated by water reuse in such industries as pulp and paper manufacture. Hardness and pH can render products inactive unless steps are taken to correct the problem, such as the inclusion of chelating agents in the formula.

In the remainder of this section, we discuss problems that are prevalent with liquid formulations, which come in several forms:

- Solutions, aqueous or otherwise
- Emulsions, with water the continuous phase, and "inverts," where oil is the continuous phase
- Slurries

10.3.1   Solutions

10.3.1.1   Drop-Out

Chemicals placed in solution can drop out of solution for a number of reasons. The most obvious is that the saturation limit has been reached and small temperature changes cause crystallization. In that case concentration must be lowered. A fairly subtle aspect of crystallization occurs when anhydrous salt is added, which, in solution, may slowly form a stable hydrate, which then crystallizes from what has become a supersaturated solution. Sodium metasilicate, for example, can form several hydrates with differing water solubilities.

When wetting agent drops out as the temperature is lowered, suspect structure. For example, if the wetting agent chosen is a straight-chain ethoxylate, change to a branched-chain surfactant with the same hydrophile–lipophile balance (HLB) or one that has a double bond in the chain. The lower melting points of the nonstraight-chain wetting agents give them better characteristics for avoiding drop-out.

A severe class of drop-out due to complex formation occurs when ingredients of the formulation react together, often nonstoichiometrically. This should always be suspected when a formula contains both cationic and anionic organic compounds. There is also the possibility of anionic species reacting with each other. Examples of both kinds of reaction are:

*   Carboxylic acids and salts of heavy metals
*   Anionic surfactants combined with quaternary N compounds
*   Silicates and borates or aluminum compounds
*   Silicates and chromates or molybdates
*   Anionic polymers and cationic polymers

The obvious answer to such problems is to replace one or the other of the suspect chemicals with one that will not react to cause drop-out. The list above is obviously not all-inclusive but will alert the formulator to the possibilities.

10.3.1.2   Splits

Why should a solution separate into two or more layers? Usually because the system is more or less colloidal, only appearing to be a solution. The first step in analyzing a solution that separates should be to determine whether it is indeed colloidal. Take a freshly prepared batch, unsplit, and shine a narrow light beam through the liquid. If you observe Tyndall scattering, the system has colloidal properties.

Several approaches can be offered to solve a problem such as this:

*   Increase the concentration and/or raise the HLB of the surfactant.
*   Use a coupling agent or hydrotrope. Simple compounds will often work, such as urea, a Cellosolve or xylene sulfonate. Otherwise, more sophisticated hydrotropes are available such as phosphate esters. Glycols may work. Propylene glycol, hexylene glycol and low molecular weight poly(ethylene glycols) are favorites.

Check for effectiveness initially by centrifugation, then by longer term storage.

10.3.1.3   Corrosion

Aqueous solutions, especially those containing dissolved salts, will corrode metals. Steel corrosion is the most prevalent, followed by the "white" metals such as aluminum, magnesium, or zinc, and their alloys [1,2].

10.3.1.3.1   Iron Corrosion

As mentioned in Section 10.2.4, iron corrodes least at about a pH of 8.5, and this property offers the first opportunity to prevent such deterioration. Several inexpensive compounds can also be used as corrosion inhibitors:

- Silicates (water glass)
- Long-chain amines
- Carboxylic acids, including polyacrylic acids
- Polyphosphates at low concentrations (ppm range)
- Alkanolamines
- Borates
- Oxygen scavengers such as nitrite or sulfite.

All are effective in alkaline media.

To inhibit acid corrosion, more sophisticated measures are taken. Among the compounds that are effective are:

- Poly(methylene imines)
- Acetylenics, such as propargyl alcohol and hexynediol
- Substituted anilines

10.3.1.3.2   White Metals and Alloys

Aluminum and zinc are amphoteric; that is, they can react (corrode) in both acid and base. Their alloys will also be attacked, dezincification of brass being an example. Magnesium/aluminum alloys are fairly resistant to corrosion except by strongly acidic or alkaline media. Anodic coatings will be attacked.

The best approach for any of these metals is to adjust the pH to a value near 7, to prevent corrosion in the first place. Compounds that can be added generally function by reacting with the surface to form an insoluble film. These include silicates, long-chain amines, and phosphates.

10.3.1.3.3   Copper and Its Alloys

By far the most effective way to prevent copper corrosion is by using compounds containing organic sulfur and nitrogen. This should come as no surprise, considering the chemistry of copper. Copper has a high affinity for both S and N, forming sulfide or amine complexes readily. Hence, compounds having these elements can be expected to chemisorb on copper surfaces, yielding stable films. Mercaptobenzothiazole is the prototype. Today benzotriazole and tolyltriazole are used more widely.

10.3.1.3.4 Glass

Glass will corrode (i.e., dissolve or etch) if the medium is too alkaline. Sodium metasilicate is used commonly as an inhibitor and may function by a mass action effect; namely, the silicate in solution prevents solubilization of the silica in the glass structure. For soft glasses, this is by no means foolproof.

For automatic dishwashing, the protection of the overglaze on fine china presents the same situation as for glass. Here, too, the washing compound contains silicate, often a metasilicate. Otherwise the alkalinity of the medium will be corrosive to the glaze. A typical compound will contain 5% or more of anhydrous metasilicate; higher amounts will be present if the pentahydrate is used.

10.3.2   Emulsions

Emulsions are more or less stable suspensions of liquid droplets in a second liquid. The two liquids are mutually immiscible. Usually the suspended phase is an oil, the continuous phase being water. An invert emulsion has water suspended in an oil phase. Emulsions are rendered stable by surfactants, which form a protective layer about the suspended droplets. It is the practice to refer to oil-in-water emulsions as o/w and water-in-oil emulsions as w/o. For further information on the basis of emulsions, see Chapter 1.

Most emulsion problems involve both long- and short-term stability. Emulsions split, cream, and do other nasty things. Production problems, which also are common, often are associated with mixing and temperature control. Sections 10.3.2.1–10.3.2.3 seek to address these problems in a practical manner.

10.3.2.1   Stability

When emulsions split or cream, the first approach, assuming that the HLB requirement of the internal oil phase has been determined, should be to examine the emulsifier system [3]. Having the HLB requirement in hand, it is not enough to have added a surfactant with that HLB. The following should be tried:

1. Bracket the HLB. That is, take two emulsifiers, one lower in HLB and the other higher than the HLB required. The amounts used should be arbitrarily set close together to permit arrival at the desired HLB value by calculating the weighted average of the two.
2. If stability is still poor, change emulsifier types. For example, from an aromatic ethoxylate to a paraffinic ethoxylate or to a sorbitan derivative. Do maintain the HLB needed.
3. If fatty acids are present in the oil phase, add some potassium hydroxide or carbonate to the system in less than a stoichiometric amount. The intent here is to form a small amount of soap, as an additional emulsifier.
4. Examine pH. Often higher pH will help in itself. A good method to destabilize emulsions is by using an "acid split," hence the view that higher pH often helps to enhance stability.
5. Try adding the ingredients in the reverse order. Usually the oil phase is introduced into the water to form o/w emulsions. The opposite may work. This means that the system will first be w/o, then will invert to o/w. As the inversion point is reached, the viscosity will

increase. Past that point, the viscosity will decrease to a workable value as the emulsion inverts. Using this technique, it is often possible to form o/w emulsions that have greater stability than is obtainable by the direct route.

6. Look for foam. When working with emulsions foam is not always easily detected, since foam has opacity and whiteness similar to most emulsions. Foam and entrained air do lead to volume increase, a useful confirmatory observation.

A few minutes in an ultrasonic bath should release foam and entrained air bubbles, which will rise to the top, becoming obvious. Foam lends temporary stability to emulsions but will eventually contribute to creaming. Foaming can also lead to manufacturing problems, tending to bind pumps or to cause vessels to overflow. To prevent foam, make sure that batch mixing is slow. Adding a defoamer is usually not productive because the active constituents of defoamers are not water soluble and will eventually rise to the top, forming a float on the surface.

Should it prove necessary to add a defoamer, avoid oil-based defoamers and silicones. Low HLB ethoxylates and phosphate esters offer the best approach, since they may form mixed micelles with the other surfactants present, avoiding the "float" problem. These will also present the least interference with detergency, a problem associated with classical defoamers that contain oils or insoluble esters. So-called water-based defoamers are emulsified long-chain alcohols, fatty acids, or waxes. These can be successful, but care must be taken to choose a defoamer that is compatible. Long-chain alcohols will function as defoamers (C_8 and higher). These are also water insoluble but may become incorporated in surfactant micelles, thus not contributing to the float problem. Simple foam testing needs to be augmented with stability studies [4,5].

In general, surfactants with terminal OH groups will tend to foam more than those in which the OH group has been "capped" (i.e., the H has been substituted by an alkyl group, usually methyl). Undue foaming can be avoided by using the capped versions. Capping can lower HLB, however, so that value should be checked if the capping has been done in-house. If the surfactant has been purchased and the HLB value determined by the supplier, compatibility with the other formula ingredients should be checked to ensure that the capping has not resulted in instability.

10.3.2.2 Viscosity

The desired emulsion viscosity is a property that depends on the application. For cosmetics, pastes are often desirable, whereas viscosities close to water are best for emulsions that are to be pumped.

When a paste is wanted, but the formulation produces a thin liquid, the problem is not serious. All that is needed is a thickener (guar gum, carboxymethylcellulose, a polyacrylate, etc.). Should these be undesirable for any reason, many powders will serve the purpose, such as finely divided silica or clay.

When the formula produces a paste but an easily pumpable liquid is wanted, ingenuity comes into play. For emulsions having water as the continuous phase, there are remedies. Most often polyfunctional molecules are useful, such as:

• Cellosolves
• Aminomethyl propanol (AMP)
• Urea

- Lignosulfonates
- Butyl carbitol
- Propylene glycol and its monoesters

and a host of low molecular weight alcohols, ketones, and esters. In some cases, the addition of surfactants will lower viscosity. Among nonionics, high HLB is often effective. Among anionics, sulfonates and phosphates may work. In some cases, the simple addition of minor amounts of alkali can lower viscosity significantly; potassium hydroxide and ethanolamines should be tested. The formulator should always keep dropping bottles available for quick-and-dirty tests. The results are poorly predictable when rheology is the issue, and the Edisonian approach is what is left.

Formulation stability must be tested following this type of adjustment. All too often the viscosity will climb again or the product will split. The chances for success in this unpredictable area are reasonable, however, and one can save a good deal of time, compared to the alternative course of restudying the entire emulsion.

10.3.2.3   Manufacturing Emulsions

The problems pointed out in the preceding section translate into manufacturing problems.

1. Temperature control must be strict. Some systems are stabilized by high temperature, some the reverse. The latter is particularly true when nonionic surfactants, which cease being emulsifiers above the cloud point, are used.
2. Mixing conditions must be appropriate. Usually low-speed mixing is preferred. For one thing, a low speed introduces less air, therefore less foam. For w/o emulsions or for systems undergoing inversion, viscosity tends to be high. A high-energy mixer will tend to drill a hole in the emulsion instead of providing efficient mixing of the entire mass.
3. Some emulsions require high energy input. This does not mean high-speed mixing but rather equipment designed to provide the energy required. Recognize that such equipment (e.g., a colloid mill) treats a small portion at a time. Therefore, ensure that the feedstock is uniform by premixing immediately before the mill or by metering the oil and water phases separately in the correct proportions directly into the mill.

10.4   Slurries

It is often necessary to formulate slurries, rather than solutions or emulsions. This is an attractive route to providing concentrated products, where the desired amount of key ingredients exceeds their solubility. Once diluted, they will dissolve rapidly to do their job.

Special problems arise when products are prepared as slurries that are then dried for packaging and shipment. The classical case is spray-dried household detergents. It was learned early on that many of these slurries would corrode plant equipment—the conveyors, spray dryers, and pumps. Other drying and agglomeration methods entail the same problem. It was found, however, that the inclusion of a few percent of sodium silicate will prevent this problem. See Chapter 6 for additional information on this topic.

The main problem with slurries is the need for uniform distribution of the chemicals and therefore the prevention of settling. No one likes containers with the legend "shake before using," particularly when the container is a drum. The most attractive course of action is to increase the viscosity sufficiently to prevent settling.

10.4.1   Viscosity Increase

Polymers of various types will increase the viscosity of water substantially at low concentrations. Among these are:

- Carboxymethylcellulose (CMC), available in various modifications as to degree of substitution and molecular weight, is generally used between 0.5 and 2%, depending on the type chosen. Carboxyethylcellulose (CEC), its cousin, works similarly. In terms of detergency, CMC or CEC additionally provides soil suspension properties and also prevents redeposition and calcium carbonate scale formation. Thus, CMC or CEC can provide additional functionality to the product, besides having raised viscosity to stabilize the slurry.
- Alginates belong to a group of naturally derived polysaccharides, obtained by the processing of seaweed. For viscosity increase, more than 1% seldom is needed. Many grades of this product are available for use in toothpaste, ice cream, and so on.
- Guar gums fall roughly into the same class as alginates for viscosity increase. Here, too, many grades are available and they are very effective, 1% or less being sufficient in most cases.
- Polyacrylic acid, having a molecular weight above 200,000 and preferably in the million range, has excellent suspension properties as well as the ability to raise viscosity. Again, many grades are available for testing. This polymer, like CMC, also acts as a soil suspending agent and a scale preventer. Chapter 5 has much detailed information on the benefits of polymers in these systems.

A frequent problem encountered when using viscosity enhancers is the control of the small amounts needed, particularly when the batch size is low. These agents should always be added gradually to the vortex of a well-stirred vessel or introduced slowly via an eductor through which the batch is circulated. Enough time must be allowed for the viscosity to reach its maximum value, with slow mixing at low shear rate. High shear mixing can break down the polymer, resulting in too low a viscosity increase.

Liquids other than water can be used for products, particularly when some of the ingredients have limited water solubility. This situation arises with polyphosphates, silicates, and borates, for example, when they are used as builders. The main thing is to keep the solids in a stable suspension, which calls for a fluid having a higher viscosity than water. Furthermore, a polymeric fluid, which exhibits non-Newtonian behavior, offers additional benefits through hindered settling. Any fluid chosen should be miscible with water and free from interference in the laundering process.

One class of polymeric liquid that has the desired properties for many such applications is poly(ethylene glycol) (PEG). PEG is available over a wide range of molecular weights, the most useful being 400–1000. Higher molecular weight PEGs are solid, but miscible with water. They can be used in conjunction with water, whereas the lower molecular weight PEGs can be used alone.

10.5   Quality Assurance

This chapter has, so far, been concerned with specific areas of concern to the formulator. In a broader sense, all the topics that have been discussed deal, in one way or another, with product quality. Quality assurance—making sure that the product that eventually arrives in the hands of the customer meets expectations—is the key to success.

The expectations of the customer are many. Certainly, the product must perform. It should be pleasant to use, as well: its physical form should be appropriate for the equipment in the home or in industry. Certainly for consumer use, it should look and smell good. The product must be properly labeled, pointing out any safety problems. In short, the product needs to be market oriented.

The market has changed over the years. Gone are the days when a small number of suppliers could control it, if that was ever really the case. Trends that began in the 1960s, of environmentalism and consumerism, rule the market. Spurred by market forces, wider environmental concerns (the Green Revolution), the media, and economic necessity, the formulator is pressed to increase involvement with customer demands and perceptions.

Identifiable trends that will continue for the foreseeable future are as follows:

Concentrates: The customer is insisting on concentrated products, to lower the amount of water that is shipped. This trend, begun in industrial and institutional applications, has affected the household market, particularly for laundry products. The old criterion of "a cupful of detergent per load" is fast disappearing.

Packaging: Packaging is being reduced in size. The trend toward concentrates is abetting this, of course. Beyond that, the consumer is increasingly conscious of wasted packaging materials and their effect on landfills. Package biodegradability and recyclability are issues that have been brought home to the consumer by both government and the media, forcing further moves to reduce and modify packages. In the background is also the increased cost and decreased availability of shelf space in retail stores.

Air Quality: Customers have become sensitized toward this issue. The VOC regulations resulted from public pressure and from "chemophobia." This trend, which began with paints and fuels, now affects cleaners. Chlorinated solvents have been under a cloud for some time; others are following.

Water Quality: Biodegradability and eutrophication are the forces driving a long-standing concern with water quality. The survival of wildlife and the contamination of fish are now part of this issue.

The trends just listed are collectively referred to as the Green Revolution, which has gathered steam in both Europe and North America. Along with increased consumer consciousness and activism, these trends have given new meaning to quality in terms of customer needs [6,7]. These concerns are real, and the formulator must account for them in product development.

10.5.1   Ingredients

A cardinal sin committed by every formulator is having too many ingredients. Product modification aimed at specific deficiencies is most often at fault. More ingredients mean greater chances for error when manufacturing the product, as well as more chances for incompatibility among the components. Products need to be reviewed, eliminating chemicals that do not meet the performance needs or overlap each other in functionality. Look for opportunities to simplify products.

Ensure that all the chemicals used conform to legal requirements, including FDA clearance; toxicity and environmental suitability are particularly important. Then reexamine the product that has met these requirements and look for chemical reactions that might occur, that may lead to such problems.

Even innocuous materials may lead to problems. Volatile solvents have fairly distinctive and pleasant odors, but combinations of solvents or solvents and scents may take on obnoxious smells. The human nose and mind can react strangely at times. The person working to develop a product may have lost sensitivity to an odor that will be offensive to a stranger. An odor panel should be part of the product development team.

10.5.2   Testing

Quality assurance depends on two main test classifications:

- Laboratory tests to ensure that the product meets specifications and
- Performance tests to ensure that the product meets customer needs.

Often these categories can be combined, but the two needs must always be kept in mind. Probably the best source for such tests is the ASTM. For cleaners, the ASTM D-12 Committee have labored long and hard to develop a body of tests covering both needs [8]. Familiarity with this work is essential.

10.6   Getting the Product out of the Laboratory

Based on the benchwork done, the formulator should be in a position to recommend the manufacturing method for a product. This implies:

- A good knowledge of manufacturing methods available. Will the product be made in batches or continuously? What is the equipment like, and what alternatives exist? Will a capital expenditure be needed? A sound knowledge here will help ensure quality and overall success.
- The ability to extrapolate from laboratory methods to larger amounts (i.e., scale-up). Plant mixing conditions will not mimic the laboratory. Both heating and cooling take longer when the total mass has been increased. Longer residence time at a high temperature may lead to unwanted side reactions such as dehydration or polymerization.
- Dosing of the proper amounts can be a problem. Production circumstances may dictate addition by weighing or by automatic metering. Neither is a problem except when very low amounts of critical ingredients need to be added. Special care must be prescribed in those cases.
- Develop the ability to communicate to others having a different background, be they engineers or operators. This does not simply mean a decent command of the language, but the willingness and the ability to express ideas in terms that the listener will understand in the context of his or her experience and training.

- Provide a valid product description, which includes physical properties, performance data, and any feature that can result in benefits to the user. The term "user" assumes its broadest meaning: internally, the user includes marketing and the sales department; externally, the user is not just the buyer of the product but also the person who eventually applies it. Even more broadly, user also includes society as a whole, in the context of safety and environmental requirements.

The foregoing concepts apply equally to technical service activities, whether by telephone or actual visits to the customer. The chemist who formulated the product is looked on as the expert in its proper application and so must be prepared to answer many questions and be able to a recommend application methods. A good knowledge of the user's circumstances is needed, whether the use be industrial, institutional, or home.

References

1. Nathan, C.C., Ed., *Corrosion Inhibitors,* National Association of Corrosion Engineers, 1973.
2. Fontana, M.G., Staehle, R.W., Eds., *Corrosion Science and Technology,* Plenum Press, New York, 1970.
3. Becher, P., *Emulsions,* American Chemical Society Monograph Series, Reinhold, New York, 1965.
4. Rosen, M.J., *Surfactants and Interfacial Technology,* Wiley, New York, 1978.
5. Linfield, W.M., Ed., Surface Sc. Ser., No. 7, Dekker, New York, 1982.
6. *Quality Assurance for the Chemical and Process Industries,* American Society for Quality Control, 1988.
7. Dixon, G., Swiler, J., *Total Quality Handbook,* Lakewood Books, Minneapolis, MN, 1990.
8. American Society for Testing and Materials, *Annual Book of ASTM Standards,* ASTM, Philadelphia.

Additional Reading

Deming, W.E., *Quality, Productivity and Comprehensive Position*, MIT Press, Cambridge, MA, 1982.

Kirk–Othmer Encyclopedia of Chemical Technology, 3rd ed.,Wiley, New York, 1978.

Schmitt, T.M., *Analysis of Surfactants,* Dekker, New York, 1991.

APPENDIX:

Suppliers of Raw Materials

The Appendix consists of two sections: a compilation of suppliers of key ingredients for detergent formulating and a glossary.

The user should already be aware that the supplier director is, almost by definition, incomplete and inaccurate. This is because we are dealing with a moving target. Individual raw materials and, indeed, entire product lines, are sold or traded among companies. Companies change their names and locations. Telephone or FAX numbers are changed as a result of this activity or at the whim of a utility. So far as possible, 800 numbers have been sought out to avoid toll call charges; where a company lists several numbers, depending on the product line, the most appropriate one has been selected.

The glossary is not intended to give dictionary definitions. Rather, the meanings are given in the context of detergency in general, and their use in these chapters in particular. In other words, they are meant to be useful, not definitive.

1. Suppliers of Surfactants

Surfactant class	Suppliers	Trade names
Alkanolamides	Alcolac	Cyclomide
	Croda	Incromide
	Henkel	Comperlan
		Emid
		Nitrene
		Standamid
	Mona	Monamid
		Monamine
	PPG/Mazer	Mazamide
	Rhône-Poulenc	Alkamide
	Scher	Schercomid
	Sherex	Varamide
	Stepan	Ninol
	Witco	Witcamide
Alkylarylsulfonates	Alcolac	Siponate
	Emkay	Emkane
	Stepan	Nacconol
		Ninate
	Witco	Hybase
		Petronate
		Witconate
Alkylpolyglycoside	Henkel	Glucopon

(continued)

1. Suppliers of Surfactants (cont'd)

Surfactant class	Suppliers	Trade names
Amine oxides	Alcolac	Cyclomax
Amine oxides	Continental	Conco
	Henkel	Standamox
	Scher	Shercamox
	Sherex	Varox
	Stepan	Ammonyx
Amines/amides, sulfonated	Emkay	Emkapon
	Mona	Monamine
	Sandoz	Sandopan
Betaines	Alcolac	Cycloteric
	Henkel	Velvetax
	Inolex	Lexaine
	Mona	Monateric
	Rhône-Poulenc	Mirataine
	Scher	Schercotaine
	Sherex	Varion
	Witco	Emcol
Block polymers	BASF-Wyandotte	Pluronic
		Tetronic
	Hart	Hartopol
	Henkel	Monolan
	Olin	Poly-Tergent
	PPG/Mazer	Macol
	Rhône-Poulenc	Antarox
		Pegol
Carboxylated alcohol - ethoxylates	Rhône-Poulenc	Miranate
	Sandoz	Sandopan
	Witco	Emcol
Diphenyl sulfonates	Continental	Conco
	Dow Chemical	Dowfax
	Olin	Poly-Tergent
Ethoxylated alcohols	Alcolac	Siponic
	BASF-Wyandotte	Plurafac
	Chemax	Chemal
	Ethyl	Ethonic
	Henkel	Dehydol
		Emery
		Eumulgin

(continued)

1. Suppliers of Surfactants (cont'd)

Surfactant class	Suppliers	Trade names
Ethoxylated alcohols	Henkel	Generol
		Sulfotex
		Trycol
	ICI-Americas	Brij
		Renex
		Synthrapol
	PPG/Mazer	Macol
		Antarox
	Rhône-Poulenc	Antarox
		Emulphogene
	Shell	Neodol
	Sherex	Arosurf
		Varonic
	Stepan	Cedepal
	Union Carbide	Tergitol
		Triton
	Vista	Alfonic
	Witco	Witconol
Ethoxylated alkyl phenols	Alcolac	Siponic
	BASF-Wyandotte	Iconol
	Desoto	Armul
		DeSonic
	Henkel	Hyonic
		Standapon
	Hoechst	Emulsogen
	Monsanto	Sterox
Ethoxylated alkyl phenols	PPG/Mazer	Avanel
		Macol
	Rhône-Poulenc	Igepal
	Texaco	Surfonic
	Union Carbide	Triton
Ethoxylated amines, amides	Chemax	Chemeen
	PPG/Mazer	Mezamide
		Mezeen
	Rhône-Poulenc	Katapol
	Sherex	Varonic
Ethoxylated fatty acids	Chemax	Chemax
	Henkel	Emerest
		Nopacol
	Calgene	Hodag

(continued)

1. Suppliers of Surfactants (cont'd)

Surfactant class	Suppliers	Trade names
Ethoxylated fatty acids	Lipo	Lipopeg
	PPG/Mazer	Mapeg
	Rhône-Poulenc	Emulphor
Ethoxylated esters, oils	Heterene	Hetoxide
	ICI-Americas	Arlatone
		Atlox
		Synthrapol
	Lonza	Glocosperse
		Lozest
	Rhône-Poulenc	Emulphor
	Sherex	Varonic
Fluorocarbons	MMM	Fluorad
Imidazolines and	Henkel	Texamine
derivatives	Lonza	Amphoterge
		Unamine
	Mona	Monaterge
		Monateric
		Monazoline
	Rhône-Poulenc	Miramine
		Miranol
		Mirataine
	Scher	Shercoteric
	Sherex	Varine
		Varion
Olefin sulfonates	Alcolac	Siponate
	Stepan	Bio-Terge
		Polystep
		Stepantan
	Witco	Witconate
Phosphate esters	Alcolac	Cyclophos
	DeSoto	DeSophos
	Harcros	T-Mulz
	Henkel	Forlanit
		Fosfamide
Phosphate esters		Fosterge
	Hoechst	Hostaphat
	Rhône-Poulenc	Antara
		Gafac
		Rhodafac
	PPG/Mazer	Maphos

(continued)

1. Suppliers of Surfactants (cont'd)

Surfactant class	Suppliers	Trade names
Phostphate esters	Union Carbide	Triton
Poly(ethylene glycol) and glycol esters	Henkel (Emery Div.)	Emerest
	Humko	Kemester
	Stepan	Kessco
	Witco	Witconol
Quaternaries	High Point	Hipochem
	Sherex	Adogen
		Variquat
	Witco	Emcol
Silicones	Dow Corning	Dow Corning
Soaps	Hercules	Dresinate
Sorbitans	ICI-Americas	Atlox
		Atsurf
		Span
		Tween
	Lipo	Liposorb
	PPG/Mazer	T-Maz
Sulfates/alcohols	Alcolac	Sipex
		Sipon
	Du Pont	Duponol
	Continental	Conco Sulfate
	Henkel	Avirol
		Standapol
		Sulfotex
		Sulfopon
		Texapon
	Stepan	Polystep
		Stepanol
	Witco	Witcolate
Sulfated ethoxylated alcohols	Exxon	Surflo
	Continental	Conco Sulfate
	Henkel	Avirol
		Standapol
		Sulfotex
		Sulfopon
		Texapon
	Stepan	Polystep
		Steol

(continued)

1. Suppliers of Surfactants (cont'd)

Surfactant class	Suppliers	Trade names
Sulfated ethoxylated alcohols	Witco	Alfonic Witcolate
Sulfonated aromatics	Ciba-Geigy	Erional Irgasol Mitin
	Continental	Conco
	DeSoto	Morwet Petro
	Henkel	Sellogen
	Stepan	Polystep Stepantan
Sulfonated aromatics	Tennessee	*p*-Toluene Sulfonic Sul-fon-ate
	Vista	Vista
	Witco	Witconate
Sulfosuccinates	American Cyanamide	Aerosol
	Mona	Monamate Monawet
	Scher	Shercopol
	Sherex	Varsulf
	Witco	Emcol

2. Suppliers of Builders

Inorganic builders	Suppliers	Trade names
Borates	Ashland Borichem Kerr-McGee U.S. Borax & Chemicals Expanded Products McKean-Rohco	Borax Exbor Borax, puffed
Phosphates	American International Chemical Ashland Chemical FMC Corporation Monsanto Occidental Olin Rhône-Poulenc	Trisodium phosphate
Polyphosphates	Albright and Wilson	

(continued)

2. Suppliers of Builders (cont'd)

Inorganic builders	Suppliers	Trade names
Polyphosphates	Browning FMC Corporation Monsanto NuSource Occidental Olin Rhône-Poulenc	
Silicates	Ashland Chemical Products Occidental PQ Corporation	Tex-Sil, Chem-Silate Metso, Kasil (potassium)
Sodium and potassium carbonates	Ashland Akzo Church and Dwight FMC Corporation General Hill Brothers Ker-McGee Occidental Rhône-Poulenc	 Trona Lite

Organic builders	Suppliers	Trade names
Citric acid	Allen Chemical American International Chemical Browning Pfizer	 Citrosol
EDTA	Ashland BASF Ciba-Geigy Grace Rhône-Poulenc Vinings	Versene Trilon Sequestrene Hampene Questex Vinkeel
NTA	BASF Grace	Trilon Hampshire NTA

3. Suppliers of Bleach

Bleach	Supplier
Chlorine	Akzo
	Atochem NA
	Georgia Gulf
	Georgia Pacific
	Kaiser Chemical
	Occidental
	Olin
Chlorinated isocyanurate	Monsanto
Hydrogen peroxide	Ashland
	Atochem NA
	Browning
	Degussa
	Du Pont
	FMC Corporation
	Solvay Interox
Sodium hydrosulfite	Ashland
	Browning
	Hoechst
	Olin
	Tennessee
	Wright
Sodium perborate	Ashland
	Atochem NA
	Browning
	Degussa
	Du Pont
	Solvay Interox
Sodium percarbonate	Browning
	Degussa
	Solvay Interox
Sodium persulfate	Ashland
	Degussa
	FMC Corporation
	Spectrum
Trisodium phosphate, chlorinated	Olin

4. Suppliers of Polymers

Polymer type	Supplier	Trade names
Alginates	Kelco Div., Merck	Kelgin
Carboxyalkylcellulose	Aqualon Div., Hercules	CMC
		Natrosol
		Cellulose gum
	Dow Chemical	Ethocell
	Union Carbide	Cellosize
Napthalene/formaldehyde	Rhône-Poulenc	Blancol
Polyacrylates	Alco Div., National Starch	Alcogum
	American Cyanamid	Cyanamer
	Colloids Inc.	Colloid
	Goodrich	Carbopol
		Goodrite
	Rohm & Haas	Acumer
Poly(vinylpyrrolidone)	GAF	Polyclar
		PVP
Styrene/maleic acid	Atochem	SMA Resin
Vinyl ether/maleic acid	GAF	Gantrez
Xanthan gum	Kelco Div., Merck	Kelzan

5. Directory of Suppliers

Company	Telephone	Telefax
Akzo	312-906-7500	312-906-7680
Albright and Wilson	804-550-4300	804-752-6185
Alco Div., National Starch	615-629-1405	
Alcolac	301-355-2600	
Allan	201-592-8122	201-592-9298
American Cyanamid	201-831-2000	
American International Chemical	800-238-0001	508-655-0927
Aqualon	800-345-8104	
Atochem	215-587-7000	215-587-7497
BASF-Wyandotte	201-316-3000	201-402-7924
BF Goodrich	800-331-1144	216-447-6392
Borichem	800-223-0662	201-299-7933
Browning	914-686-0300	914-686-0310
Calgene	800-432-7187	
Chemax	800-334-6234	803-277-7807

(continued)

5. Directory of Suppliers (cont'd)

Company	Telephone	Telefax
Chemical Products	404-382-2144	
Church and Dwight	609-683-5900	
Ciba-Geigy	800-334-9481	919-632-7008
Continental	201-472-5000	201-472-5221
Croda	212-683-3089	
Degussa	201-641-6100	201-807-3183
DeSoto	817-847-4400	817-847-4444
Dow Chemical	800-447-4369	
Dow Corning	517-496-4000	
Du Pont	800-441-7515	302-774-7573
Emkay	201-352-7053	
Ethyl	800-535-3030	504-388-7686
Expanded Products	201-839-4022	
Exxon	800-526-0749	713-870-6661
FMC Corporation	215-299-6000	215-299-5999
GAF Corporation	201-628-3000	
General	800-631-8050	
Georgia Pacific	404-521-4000	404-827-7010
Grace	301-659-9000	301-685-1434
Harcros	913-321-3131	913-621-7718
Hart	717-668-0001	717-668-6526
Henkel	800-531-0815	215-628-1353
Hercules	800-247-4372	
Heterene	201-278-2000	201-278-7512
High Point	919-884-2214	919-884-5039
Hill Brothers	714-998-8800	714-998-6310
Hoechst	804-393-3100	804-393-3246
Humko	901-320-5800	901-682-6531
ICI-Americas	800-634-8309	302-652-8836
Inolex	800-521-9891	215-289-9065
Kaiser Chemicals	216-292-9226	
Kerr-McGee	800-654-3911	
Lipo	201-345-8600	201-345-8365
Lonza	201-794-2400	201-794-2597
MMM	612-736-1394	
McKean-Rohco	216-441-4900	
Merck	201-574-4000	
Mona	201-345-8220	
Monsanto	800-325-4330	314-694-7625
National Starch	201-685-5000	201-685-5005

(continued)

5. Directory of Suppliers (cont'd)

Company	Telephone	Telefax
NuSource	415-433-2223	415-433-2526
Occidental	716-696-6000	716-696-6260
Olin	203-356-2000	
Pfizer	201-470-7721	201-470-7877
PPG/Mazer	708-244-3410	708-244-9633
PQ Corporation	215-293-7200	215-688-3835
Rhône-Poulenc	609-395-8300	201-297-3500
Rohm & Haas	215-592-3000	
Sandoz	704-331-7000	704-377-1064
Scher	201-471-1300	201-471-3783
Sherex	800-366-6500	614-764-6544
Solvay-Interox	800-468-3769	713-524-9032
Spectrum	800-772-8786	213-516-9843
Stepan	312-446-7500	312-501-2443
Tennessee	404-239-6700	404-239-6701
Texaco	713-961-3711	
U.S. Borax & Chemical	800-872-6729	213-251-5455
Union Carbide	800-243-8160	
Vinings	800-347-1542	404-436-3432
Vista	713-588-3000	713-588-3236
Witco	212-605-3981	201-605-3660
Wright	919-251-8952	919-762-9223

Glossary

ABS	Alkylbenzene sulfonate
Adhesion	Strong attraction between two materials resulting in binding.
Agglomeration	Binding of particles into larger masses.
AES	Alcohol ether sulfate wetting agent.
Alcohol ether sulfate	Sulfated, ethoxylated alcohol wetting agent.
Alcohol ethoxylate	Ethoxylated long-chain alcohol wetting agent.
Alkylphenol ethoxylate	Nonionic wetting agent from ethoxylated alkylate.
Alkane sulfonate	Sulfonated paraffin wetting agent.
Alkyl sulfate	Alcohol sulfate wetting agent.
Alkylbenzenesulfonate	Synthetic anionic wetting agent.
Alkyl polyglycoside	Corn-derived wetting agent.
α-Olefin sulfonate	Sulfonated olefin wetting agent.
Amphiphile	Molecule having two parts, one hydrophilic, the other lipophilic.
Amphoteric surfactant	Surfactant having both negatively and positively charged functional groups.
Amylase	See *Enzyme*.
Anionic surfactant	Negatively charged surfactant molecule (e.g., LAS).
Antistat	Additive that eliminates static electrical charge.
APE	Alkylphenol ethoxylate
Bleach	Oxidizing agent that reacts with chromopores, causing color change toward decolorization.
Bleach activator	Compounds that react with peroxygen bleaches to improve bleaching by forming percarboxylates.
Builder	A chemical that aids in soil removal without significant influence on surface tension.
CAC	Critical aggregation concentration. Applied generally to micelles (see *cmc*), liquid crystals, vesicles.
Cationic surfactant	Positively charged surfactant molecule.
Carding	Mechanical process for aligning fibers in preparation for spinning into yarn.
Chelant	Chemical that complexes metal ions in soluble form.

cmc	Critical micelle concentration.
CMC	Carboxymethylcellulose.
Coupling agent	Compound that promotes mutual compatibility of otherwise immiscible compounds. See also *Hydrotrope*.
Critical micelle concentration	Concentration at which micelles form because solubility limit has been reached.
Detergent	A formulation that will remove soil.
Emulsion	Suspension of colloidal-sized immiscible liquid droplets in another liquid; two phases present.
Enzyme	For detergency, an amino acid polymer derived from bacteria that breaks down proteins (protease), fats (lipase), or starches (amylase).
EO	Ethylene oxide, the most commonly used hydrophilic building block for nonionic surfactants.
EO/PO	Type of surfactant made using ethylene oxide and propylene oxide. PO is the hydrophobe, EO the hydrophile. May be random or structured polymer.
Fabric	Woven, knitted, or felted material composed of fibers.
Foam	Dispersion of a gas in a liquid.
Formulation	A combination of two or more chemicals.
Fluorescent whitener	Organic compound, added to wash, which absorbs ultraviolet light and emits blue light.
Glaze	See *Overglaze*.
Hardness	See *Water hardness*.
Hemimicelle	Lamellar aggregation of surface active molecules at an interface.
HLB	Hydrophile/lipophile balance.
Hydrophile	Functional group attractive to water.
Hydrophobe	Functional group compatible with oils; "water-hating."
Hydrotrope	Molecule that enhances compatibility between surfactants or between surfactants and solvent.
I&I	Industrial and institutional.
Interface	Any border between phases.
Interfacial tension	See *Surface tension*.
Inverted micelle	Micelle having polar groups in the interior of the structure.
J Box	Textile processing unit used for bleaching, scouring; shaped like the letter J.
Kier	An oversized boiler for scouring fiber using either atmospheric or modest pressure.
Krafft boundary	Surfactant solubility limit.
Lamella	Flat sheet.

LAS	Linear alkylbenzenesulfonate, a class of anionic surfactants.
Lipase	See *Enzyme.*
Lipophile	"Oil-loving" group of surfactant molecules. See *Hydrophobe.*
Liquid crystal	Ordered, usually two-dimensional, array of molecules in suspension. Some soaps form liquid crystals.
Lyotropic	Solvent induced.
Micelle	Colloidal-sized globular assembly of surfactant molecules: forms when solubility in water has been exceeded. Polar groups are on the exterior of the structure. See also *Inverted micelle.*
Microemulsion	Single-phase colloidal dispersion.
Nonionic surfactant	Molecule having no charged functional groups, but having hydrophile(s) and hydrophobe(s) in its structure.
Optical brightener	See *Fluorescent whitener.*
Overglaze	Final clear glaze applied to dishware. Must withstand alkalinity of automatic dishwashing to be effective today.
PEG	Poly(ethylene glycol).
PEG esters	Esters of PEG and fatty acids; nonionic surfactants. Diesters are low-foaming compounds that may be useful as antifoaming agents. Esters will hydrolyze.
PO	Propylene oxide.
Protease	See *Enzyme.*
Quats	Quaternary ammonium surfactants; may be ethoxylated to give nonionic character.
Redeposition	Tendency of soil removed by washing to return to the fabric during rinsing.
Rheology	Flow behavior.
Saponification	Decomposition of esters to their constituent alcohols and fatty acids. Usually accomplished with alkali.
Scour	Textile industry term for washing compound.
Sequestration	Ability of a compound to combine with metal ions in a soluble form, often nonstoichiometrically.
Soap	Salt of fatty acid; often used to designate all detergents.
Softness	Pleasant fabric hand-feel.
Soil	Any unwanted foreign material.
Substantivity	Ability of a compound to combine strongly with a fiber surface, by salt formation or by multiple hydrogen bonding.

Surface tension	Force attracting molecules at the surface to the bulk of the material or energy difference between two phases (interfacial tension).
Surfactant	A chemical that reduces the surface tension of water, mainly; can apply to other solvents.
Suspension	Stable mixture of a solid in a liquid.
Tenside	See *Surfactant*. Term used in Europe.
Thermotropic	Temperature induced.
Vesicle	Lamellar dispersion of surfactant.
VOC	Volatile organic compound.
Washing	The combination of mechanical and chemical means to remove soil.
Water hardness	Presence of heavy metal ions, commonly Ca or Mg, in solution.
Water softness	Absence of high concentrations of Ca or Mg. Water is considered soft when concentration is below 100 ppm (as $CaCo_3$).
Wetting	Ability of a liquid to spread over a surface.
Zwitterionic	See *Amphoteric*.

Index